ANNALES ACADEMIÆ SCIENTIARUM FENNICÆ

MATHEMATICA

DISSERTATIONES

132

INEQUALITIES FOR CONFORMAL CAPACITY, MODULUS, AND CONFORMAL INVARIANTS

VILLE HEIKKALA

University of Helsinki, Department of Mathematics

To be presented, with the permission of the Faculty of Science of the University of Helsinki, for public criticism in Auditorium XII, the Main Building, on December 7th, 2002, at 10 a.m.

HELSINKI 2002
SUOMALAINEN TIEDEAKATEMIA

ISSN 1239-6303
ISBN 951-41-0927-9

Received 10 October 2002

2000 Mathematics Subject Classification:
Primary 30C65; Secondary 31B15.

YLIOPISTOPAINO
HELSINKI 2002

Acknowledgements

I express my deepest gratitude to my advisor Matti Vuorinen for his guidance during the preparation of this thesis. His active but patient participation has indeed played a key role in my first steps on the path of mathematical research work.

I address my gratitude to Matti Lehtinen and Alexander Yu. Solynin for their work in the pre-examination of my work. I thank Jussi Väisälä for his ideas for the proof of Theorem 4.21 and for Remark 7.4 as well as Jan Malý for his idea which gave birth to Theorem 4.6. I am grateful to Pentti Järvi, Songliang Qiu, and Vladimir Miklyukov for reading and commenting on my manuscript. I also thank Dimitrios Betsakos and Toshiyuki Sugawa for their fruitful comments and advice.

I am indebted to the Graduate School of Mathematical Analysis and Logic in the University of Helsinki, to the Department of Mathematics of the University of Helsinki, to the Academy of Finland, and to the Finnish Society of Sciences and Letters (Magnus Ehrnroothin säätiö) for financial support.

Finally, I express my love and gratitude to Rea for her support and patience as well as to my parents Olli and Ritva, who have always supported my scientific efforts both mentally and financially.

Helsinki, November 2002

Ville Heikkala

Contents

1. Introduction

In 1950, the extremal length of a curve family was introduced by L. V. Ahlfors and A. Beurling in [AB]. Its roots lie in the length-area method, which has been widely applied in geometric function theory. In this context one usually rather considers the modulus, which is simply the reciprocal of the extremal length. Later, in 1957, B. Fuglede generalized this concept to higher dimensions, paving the way for F. W. Gehring and J. Väisälä, who in their pioneering work (see e.g. [Ge2] and [Väl]) applied the modulus method to the study of quasiconformal mappings in the 1960's.

The study of non-homeomorphic n-dimensional quasiregular mappings was initiated by Yu. G. Reshetnyak in 1966. His methods were analytic, relying on tools from differential geometry and PDE theory. Another approach to this theory, due to O. Martio, S. Rickman, and J. Väisälä, was developed in 1969–72. Their techniques were rather geometric and based on the modulus method. The basic properties of quasiregular mappings can be found in [HKM], [Re2], [Ri], and [Vu4].

There is also a variation of the modulus method, due to M. Vuorinen and his collaborators, in which the modulus is replaced by two conformal invariants λ_G and μ_G, defined as extremal moduli of certain special curve families in a domain G in $\mathbb{R}^n$. This theory has two main parts, namely the study and estimation of the above conformal invariants and their transformation rules under quasiregular mappings. It gives the advantage of a unified distortion theory for both quasiconformal and quasiregular mappings as both can be studied in the same setup. These ideas were first described and used in [Vu3] and further developed in a series of monographs and papers [Vu4], [LeVu], [Vu5], [ALV], [SV], [Fe], [AVV], [Se], [Bet], and [BV].

We are mainly involved with the first part of this theory, providing estimates for the modulus as well as λ_G and μ_G. We also study conformal capacity, a concept closely related to the modulus of a curve family. We provide new upper and lower bounds for the capacity of a condenser consisting of a compact set and its t-neighborhood for $t > 0$. Most of our results are based on thickness assumptions posed either on the set whose capacity is studied or on the boundary of G in the study of μ_G. Such assumptions are Ahlfors regularity (see [DS]), uniform perfectness (defined by Ch. Pommerenke in 1979, see the survey [Su] or [JV]), and a continuum criterion of O. Martio, defined in [Mar1] and further studied in [MS]. We also investigate the interdependence of these criteria.

Section 2 begins with a review of the preliminaries and then proceeds to the investigation of the behaviour of the modulus. It is well known (see e.g. [Vu4]) that the modulus of a curve family joining two continua E and F in $\mathbb{R}^n$ is related to the quantity $\min\{d(E), d(F)\}/d(E,F)$. We prove inequalities which clarify this relationship further. There are also some growth estimates for the modulus in Section 5.

In Section 3 we restrict our attention to the plane, where many of the quantities introduced in Section 2 have explicit representations or at least better estimates than in higher dimensions. We study the two dimensional Teichmüller capacity, the function p which provides an answer to Teichmüller's extremal problem, and the modulus.

The topic of Section 4 is the relationship between the conformal capacity and the thickness of a compact set in $\mathbb{R}^n$. We study, for compact sets E, the behaviour of the capacity $\operatorname{cap}(E+B^n(t), E)$ when $t \to 0$. Here $E + B^n(t)$ denotes the set $\{z \in \mathbb{R}^n \mid d(z, E) < t\}$. Upper and lower bounds in terms of t are obtained for Ahlfors regular and uniformly perfect sets.

In Section 5 we discuss the continuum criterion of Martio. In both Sections 4 and 5 we compare the thickness criteria, proving that Ahlfors regularity implies uniform perfectness, which furthermore guarantees that the continuum criterion holds. We also provide counterexamples to show that converse implications fail to hold.

We begin Section 6 by recalling the definitions and some basic properties of the conformal invariants λ_G and μ_G. Then we proceed to compare the topologies defined by these quantities to the Euclidean topology. We prove that if the boundary of G is sufficiently thick, namely connected or at least uniformly perfect, then the topology defined by μ_G is equivalent with the Euclidean one. For λ_G^{-1} this property is seen to hold regardless of the structure of the boundary of G. We also show that the continuum criterion on the boundary of G is a necessary assumption for the topology defined by μ_G to be equivalent with the Euclidean one.

Finally, we study how $\lambda_G(x, y)$, for fixed $x, y \in G$, changes when a point is removed from the domain G. We especially concentrate on the case when G is the unit ball in $\mathbb{R}^n$ and the point to be removed is the origin. We prove new estimates for $\lambda_{B^n\setminus\{0\}}$ and comparison results between $\lambda_{B^n\setminus\{0\}}$ and λ_{B^n}. This is the content of Section 7.

2. Preliminaries

We mainly adopt the notation and definitions of [Vu4] and [Vä2]. However, we shall review here most of the background material needed in order to make the exposition self-contained for the convenience of the reader.

2.1. **Notation and terminology.** In this paper we assume that n is a fixed integer with $n \geq 2$. We denote by $\mathbb{R}^n$ the n-dimensional Euclidean space and by $e_1, \dots, e_n$ the standard unit vectors of $\mathbb{R}^n$. For $x \in \mathbb{R}^n$, we write $x = (x_1, \dots, x_n) = x_1 e_1 + \dots + x_n e_n$. We denote the Euclidean norm by $|\ |$ and employ the abbreviations

$$B^n(x,r) = \{z \in \mathbb{R}^n \mid |z - x| < r\},\ B^n(r) = B^n(0,r),\ B^n = B^n(1),$$

$$S^{n-1}(x,r) = \{z \in \mathbb{R}^n \mid |z - x| = r\},\ S^{n-1}(r) = S^{n-1}(0,r),\ S^{n-1} = S^{n-1}(1).$$

The surface area of S^{n-1} is denoted by ω_{n-1} and Ω_n is the volume of B^n. It is well known that $\omega_{n-1} = n\Omega_n$ and that

$$\Omega_n = \frac{\pi^{n/2}}{\Gamma(1 + n/2)}$$

for all $n = 2, 3, \ldots$, where Γ denotes Euler's gamma function. By the properties of the gamma function one obtains the explicit formulae

$$\omega_{2k-1} = \frac{2\pi^k}{(k-1)!} \quad \text{and} \quad \omega_{2k} = \frac{2^{k+1}\pi^k}{1 \cdot 3 \cdots (2k-1)}$$

for $k = 1, 2, \ldots$. We will denote by m_k the k-dimensional Lebesgue measure of $\mathbb{R}^n$ and abbreviate $m = m_n$. The *Möbius space* $\overline{\mathbb{R}}^n$ is the one-point compactification $\mathbb{R}^n \cup \{\infty\}$ of $\mathbb{R}^n$. The *Poincaré half-space* in $\mathbb{R}^n$ is the set $H^n = \{x \in \mathbb{R}^n \,|\, x_n > 0\}$. For $x,\ y \in \mathbb{R}^n$, let $J_{xy} = [x, y] = \{(1-t)x + ty \,|\, 0 \le t \le 1\}$ and for $x \in \mathbb{R}^n \setminus \{0\}$, let $J_{x\infty} = [x, \infty] = \{tx \,|\, t \ge 1\} \cup \{\infty\}$. We also denote $(x, y) = \{(1-t)x + ty \,|\, 0 < t < 1\}$.

When A is a set in $\mathbb{R}^n$ or $\overline{\mathbb{R}}^n$, $\overline{A}$ means the *closure*, ∂A the *boundary* and $\complement A = \overline{\mathbb{R}}^n \setminus A$ the *complement* of A, all taken with respect to $\overline{\mathbb{R}}^n$. For nonempty subsets A and B of $\overline{\mathbb{R}}^n$, we let $d(A) = \sup\{|x-y| \,|\, x, y \in A\}$ be the *diameter* of A, denote by $d(A, B) = \inf\{|x-y| \,|\, x \in A, y \in B\}$ the Euclidean *distance* between the sets A and B, and in particular, $d(x, B) = d(\{x\}, B)$. When $x \in \overline{\mathbb{R}}^n$, we often abbreviate $d(x) = d(x, \partial A)$ if there is no danger of confusion concerning the set A. A *domain* is an open, connected, nonempty set and a *neighborhood* of a point means an open set containing it.

The set $\{0, 1, 2, \ldots\}$ of natural numbers will be denoted by $\mathbb{N}$.

2.2. **Conformal mappings.** Let D, D' be domains in $\mathbb{R}^n$. A homeomorphism $f : D \to D'$ is called *conformal* if

1. f has continuous first partial derivatives,
2. the Jacobian of f vanishes nowhere in D, and
3. $|f'(x)h| = |f'(x)||h|$ for all $x \in D$ and all $h \in \mathbb{R}^n$.

If D and D' are domains in $\overline{\mathbb{R}}^n$, then a homeomorphism $f : D \to D'$ is *conformal* if its restriction to $D \setminus \{\infty, f^{-1}(\infty)\}$ is conformal in the above sense. It is well known that the set of all conformal mappings of $\overline{\mathbb{R}}^n$ forms a group with respect to the composition of mappings. We call a quantity a *conformal invariant* if it remains unchanged under all conformal mappings.

The following transformations are examples of conformal mappings.

(1) For $t \in \mathbb{R}$ and $a \in \mathbb{R}^n \setminus \{0\}$ let $P(a, t)$ denote the hyperplane in $\overline{\mathbb{R}}^n$ perpendicular to a and at a distance $t/|a|$ from the origin, namely

$$P(a, t) = \{x \in \mathbb{R}^n \mid x \cdot a = t\} \cup \{\infty\}.$$

The reflection in $P(a, t)$ is defined by

$$x \mapsto x - 2(x \cdot a - t)\frac{a}{|a|^2}, \quad \infty \mapsto \infty.$$

(2) An inversion in the sphere $S^{n-1}(a, r)$, $a \in \mathbb{R}^n$, $r > 0$:

$$x \mapsto a + \frac{r^2(x-a)}{|x-a|^2}, \quad a \mapsto \infty, \quad \infty \mapsto a.$$

Other examples of conformal mappings are the translation $x \mapsto x+a$, $a \in \mathbb{R}^n$, $\infty \mapsto \infty$, the stretching $x \mapsto kx$, $k > 0$, $\infty \mapsto \infty$, and an orthogonal mapping, i. e. a linear mapping h with $|h(x)| = |x|$, $h(\infty) = \infty$. By [Vu4, 1.3.] and [Bea2, 3.1.3], each of these mappings can be written as a composition of a finite number of reflections in hyperplanes and inversions in spheres.

2.3. **Möbius transformations.** A homeomorphism $f : \overline{\mathbb{R}}^n \to \overline{\mathbb{R}}^n$ is called a *Möbius transformation* if there exists an integer m such that $f = f_1 \circ \cdots \circ f_m$, where each f_i, $i = 1, 2, \ldots, m$, is an inversion in a sphere or a reflection in a hyperplane. It is easy to see that Möbius transformations constitute a subgroup of the group of conformal mappings of $\overline{\mathbb{R}}^n$.

2.4. **Notation.** When E,F,$G \subset \overline{\mathbb{R}}^n$, we denote by $\Delta(E, F; G)$ the family of all non-constant curves joining E and F in G. We denote briefly $\Delta(E, F) = \Delta(E, F; \overline{\mathbb{R}}^n)$.

2.5. **Modulus and capacity.** Let Γ be a family of curves in $\overline{\mathbb{R}}^n$. We denote by $\mathcal{F}(\Gamma)$ the set of all such non-negative Borel-measurable functions $\rho : \mathbb{R}^n \to \mathbb{R} \cup \{\infty\}$ which satisfy $\int_\gamma \rho\, ds \geq 1$ for each locally rectifiable curve $\gamma \in \Gamma$. The *modulus* of Γ is

$$\mathsf{M}(\Gamma) = \inf_{\rho \in \mathcal{F}(\Gamma)} \int_{\mathbb{R}^n} \rho^n dm.$$

More generally one can define the p-modulus of a curve family and the p-capacity for $p \geq 1$ by replacing ρ^n with ρ^p in the integration above. We do not, however, consider these quantities since they fail to be conformally invariant if $p \neq n$.

By [Vä2, 6.2.], the modulus is an outer measure in the space of all curves in $\overline{\mathbb{R}}^n$. Let Γ_1 and Γ_2 be curve families in $\overline{\mathbb{R}}^n$. We say that Γ_2 is *minorized* by Γ_1, denoted $\Gamma_2 > \Gamma_1$, if every $\gamma \in \Gamma_2$ has a subcurve belonging to Γ_1. If $\Gamma_1 < \Gamma_2$, then $\mathcal{F}(\Gamma_1) \subset \mathcal{F}(\Gamma_2)$ and hence $\mathsf{M}(\Gamma_1) \geq \mathsf{M}(\Gamma_2)$. This fact will be used repeatedly in what follows.

Given compact disjoint sets $E, F \subset \overline{\mathbb{R}}^n$, there is an explicit formula for $\mathsf{M}(\Delta(E, F))$ only in very few special cases. One of these is the following formula, which was proved by Väisälä.

2.6. **Lemma.**[Vä2, 7.5.] *Let* $0 < a < b < \infty$. *Then*

$$\mathsf{M}(\Delta(S^{n-1}(a), S^{n-1}(b); \overline{B^n(b)} \setminus B^n(a))) = \omega_{n-1} \left(\log \frac{b}{a}\right)^{1-n}. \qquad \square$$

When the modulus of a curve family cannot be explicitly computed, one has to settle for estimates. The next lemma is a well known result of Väisälä, which we shall invoke repeatedly in the sequel.

2.7. **Lemma.**[Vä2, 10.12.] *Let* $0 < a < b$ *and let* E *and* F *be sets in* $\mathbb{R}^n$ *with*

$$E \cap S^{n-1}(t) \neq \emptyset \neq F \cap S^{n-1}(t)$$

for $t \in (a, b)$. *Then*

$$\mathsf{M}(\Delta(E, F; \overline{B^n(b)} \setminus B^n(a))) \geq c_n \log \frac{b}{a}.$$

Equality holds if $E = (ae_1, be_1)$ *and* $F = (-be_1, -ae_1)$. *The number* c_n, *which depends only on the dimension* n, *is defined in* [Vä2, 10.11.]. $\square$

If D is a domain in $\overline{\mathbb{R}}^n$ and $\overline{\mathbb{R}}^n \setminus D$ has exactly two components E and F, then D is termed a *ring* and denoted $D = R(E, F)$. The *(conformal) capacity* of a ring $D = R(E, F)$ is defined by

$$\mathrm{cap} R(E, F) = \mathsf{M}(\triangle(E, F))$$

and the *(conformal) modulus* of D is

$$\mathrm{mod} D = \left(\frac{\mathrm{cap} D}{\omega_{n-1}} \right)^{1/(1-n)}.$$

Let $A \subset \mathbb{R}^n$ be open and let $C \subset A$ be compact. Then the pair (A, C) is called a *condenser*. The *capacity* of a condenser (A, C) is

$$\mathrm{cap}(A, C) = \mathsf{M}(\triangle(C, \partial A; A)).$$

Note that by [Vu4, (5.10)] we have that

$$\mathsf{M}(\triangle(C, \partial A; A)) = \mathsf{M}(\triangle(C, \partial A)) = \mathsf{M}(\triangle(C, \partial A; A \setminus C)).$$

When $E \subset \mathbb{R}^n$ is a closed set, $x \in \mathbb{R}^n$ and $r > 0$, we denote

$$\mathrm{cap}(x, E, r) = \mathrm{cap}(B^n(x, 2r), \overline{B^n(x, r)} \cap E).$$

The most significant property of the modulus and the capacity is that they are invariant under conformal mappings (see [Vä2, 8.1.]). This fact makes them very useful in the study of quasiconformal and quasiregular mappings. Namely, measuring how much moduli or capacities change under a quasiconformal or quasiregular mapping tells us how much such a mapping differs from a conformal one. In fact, quasiconformal mappings are defined as follows, see [Vä2].

2.8. **Quasiconformal mappings.** Let D and D' be domains in $\overline{\mathbb{R}}^n$, $f : D \to D'$ a homeomorphism, and $K \in [1, \infty)$. Then f is *K-quasiconformal* if

$$\mathsf{M}(\Gamma)/K \le \mathsf{M}(f\Gamma) \le K\mathsf{M}(\Gamma)$$

for every curve family Γ in D, where $f\Gamma = \{f \circ \gamma \,|\, \gamma \in \Gamma\}$. A 1-quasiconformal mapping is called *conformal.* This is in accordance with definition 2.2 above by [Vä2, 8.1. and 13.7.]. For an equivalent, analytic definition of quasiconformality we refer the reader to chapter 4 of [Vä2]. For the definition of quasiregular mappings we refer to chapter 10 of [Vu4].

A compact set $E \subset \mathbb{R}^n$ is of *capacity zero*, denoted $\mathrm{cap} E = 0$, if there exists a bounded open set $A \supset E$ with $\mathrm{cap}(A, E) = 0$. A compact set $E \subsetneq \overline{\mathbb{R}}^n$ is of *capacity zero* if E can be mapped by a Möbius transformation onto a bounded set of capacity zero. Otherwise E is said to be of *positive capacity*, denoted $\mathrm{cap} E > 0$. Observe that the condition $\mathrm{cap} E > 0$ does not mean that the "capacity" of E is thought of as a real number. Countable compact sets are examples of sets of capacity zero. Sets of capacity zero are always very thin in the following sense (see 4.13 for the definition of H^α).

2.9. **Lemma.**[Re1, p. 120, Cor. 2] *Suppose that F is a compact set in $\mathbb{R}^n$ of capacity zero. Then for every $\alpha > 0$, the α-dimensional Hausdorff measure $H^\alpha(F)$ of F is zero. In particular,* $\mathrm{int} F = \emptyset$, *and F is totally disconnected.* □

As can easily be seen from definition 2.8, sets of capacity zero are preserved under quasiconformal maps.

2.10. **Spherical symmetrization.** Let $x_0 \in \mathbb{R}^n$, $E \subset \overline{\mathbb{R}}^n$ a Borel set and let L be a ray from x_0 to ∞. Then the *spherical symmetrization* E^* of E is defined as follows:

(1) $x_0 \in E^*$ if and only if $x_0 \in E$;

(2) $\infty \in E^*$ if and only if $\infty \in E$;

(3) for $r \in (0, \infty)$, $E^* \cap S^{n-1}(x_0, r) \neq \emptyset$ if and only if $E \cap S^{n-1}(x_0, r) \neq \emptyset$, in which case $E^* \cap S^{n-1}(x_0, r)$ has the same m_{n-1} measure as $E \cap S^{n-1}(x_0, r)$ and it is a closed spherical cap centered on L, or in other words, the intersection of $S^{n-1}(x_0, r)$ with a circular cone with vertex x_0 and axis along L.

Let (A, C) be a condenser and $x_0 \in \mathbb{R}^n$. Let C^* and B be the spherical symmetrizations of C and $\mathbb{R}^n \setminus A$ in two opposite rays emanating from x_0 and let $A^* = \mathbb{R}^n \setminus B$. Then (A^*, C^*) is a condenser and the following theorem holds (see [Ge1] and [Sa1]).

2.11. **Theorem.** *If (A, C) is a condenser, then*

$$\operatorname{cap}(A, C) \geq \operatorname{cap}(A^*, C^*). \qquad \square$$

The following two rings have a special role in geometric function theory. The main reason for this is that they have important extremal properties connected with spherical symmetrization of condensers, see Section 7 of [Vu4]. In the sequel we will often deduce lower bounds for the capacities of condensers by transforming the condenser at hand to one of the following rings by Möbius mappings and spherical symmetrization and then invoking Theorem 2.11.

2.12. **The Grötzsch and Teichmüller rings.** The *Grötzsch ring* in $\overline{\mathbb{R}}^n$ is

$$R_{G,n}(s) = R(\overline{B}^n, [se_1, \infty]),\ s > 1,$$

the *bounded Grötzsch ring* is

$$R^b_{G,n}(r) = R(\overline{\mathbb{R}}^n \setminus B^n, [0, re_1]),\ 0 < r < 1,$$

while the *Teichmüller ring* in $\overline{\mathbb{R}}^n$ is the ring

$$R_{T,n}(s) = R([-e_1, 0], [se_1, \infty]),\ s > 0.$$

We denote $\gamma_n(s) = \operatorname{cap} R_{G,n}(s)$ and $\tau_n(s) = \operatorname{cap} R_{T,n}(s)$. By [AVV, 8.37.], the functions γ_n and τ_n are continuous and strictly decreasing with range $(0, \infty)$. Furthermore, these functions satisfy the identity ([Vu4, 5.53.])

$$\gamma_n(s) = 2^{n-1}\tau_n(s^2 - 1),\ s > 1. \tag{2.13}$$

By conformal invariance, $\operatorname{cap} R^b_{G,n}(r) = \gamma_n(1/r)$.

It follows directly from [AVV, Theorem 8.36] that the function $t \mapsto \operatorname{mod} R_{G,n}(t) - \log t$ is positive and strictly increasing for $t > 0$. Hence it has a positive limit as $t \to \infty$ and we may define a number λ_n, the *Grötzsch ring constant*, by setting

$$\log \lambda_n = \lim_{t \to \infty} (\operatorname{mod} R_{G,n}(t) - \log t).$$

The fact that λ_n is finite is nontrivial and has been proved in [Ge1, Lemma 8] for $n = 3$ and in [Ca, Lemma 27] for higher dimensions. The only known exact value for λ_n is $\lambda_2 = 4$, which follows from [LeVi, p. 62, (2.11)]. For a summary of estimates for other values of λ_n we point the reader to [AVV, Chapter 12] and to the references therein.

In what follows we shall apply the following bounds from [Vu4, (7.27)]. Namely,

$$\max\{v_1(t), v_2(t)\} \le \gamma_n(t) \le \min\{u_1(t), u_2(t)\}, \tag{2.14}$$

where

$$u_1(t) = \omega_{n-1}\mu(1/t)^{1-n}, \qquad u_2(t) = 2^{n-1}c_n\mu\left(\frac{t-1}{t+1}\right),$$

$$v_1(t) = \omega_{n-1}(\log \lambda_n t)^{1-n}, \quad v_2(t) = 2^{n-1}c_n \log\left(\frac{t+1}{t-1}\right).$$

Here c_n is the number in Lemma 2.7 and μ is the function defined in (3.3) below. Also the following results concerning the Teichmüller capacity function are most useful for our purposes.

2.15. **Theorem.**[AVV, 11.25.] *For $n \ge 2$, $x > 0$,*

$$1 < \frac{\tau_n(cx)}{\tau_n(x)} < \frac{1}{\sqrt{c}} \quad \textit{for} \quad 0 < c < 1, \tag{2.16}$$

$$\frac{1}{\sqrt{c}} < \frac{\tau_n(cx)}{\tau_n(x)} < 1 \quad \textit{for} \quad 1 < c < \infty. \quad \square \tag{2.17}$$

2.18. **Theorem.**[AVV, 11.27.] *Let $n \ge 2$, $0 < x < \infty$. Then*

$$c \le \frac{\tau_n(x^c)}{\tau_n(x)} \le c^{1-n} \quad \textit{for} \quad 0 < c < 1. \tag{2.19}$$

$$c^{1-n} \le \frac{\tau_n(x^c)}{\tau_n(x)} \le c \quad \textit{for} \quad 1 < c < \infty. \tag{2.20}$$

$$\lim_{x\to 0} \frac{\tau_n(x^c)}{\tau_n(x)} = c, \ \lim_{x\to\infty} \frac{\tau_n(x^c)}{\tau_n(x)} = c^{1-n} \quad \textit{for each} \quad c > 0. \quad \square \tag{2.21}$$

2.22. **The extremal problems of Grötzsch and Teichmüller.** The Grötzsch and Teichmüller rings arise from extremal problems of the following type, which were first posed for the case of the plane: Among all ring domains which separate two given closed sets E_1 and E_2 with $E_1 \cap E_2 = \emptyset$, find one whose modulus has the greatest value. In [Te] Teichmüller considered the following particular case, which we will refer to as the *Teichmüller problem*: For $z \in \mathbb{C}\setminus\{0,1\}$, find the maximal modulus of all ring domains with complements consisting of two components E and F such that $0, 1 \in E$ and $z, \infty \in F$.

Let E_1 be a continuum and E_2 consist of two points not separated by E_1. By the conformal invariance of the modulus one may then suppose that $E_1 = S^1$ and $E_2 = \{0, r\}$, $0 < r < 1$. Then the extremal problem is solved by the bounded Grötzsch ring $R(\overline{\mathbb{R}^2} \setminus B^2, [0, r])$. In other words, $\operatorname{cap}(B^2, E) \ge \gamma_2(1/r)$, where $E \subset B^2$ is any continuum joining the points 0 and $r \in \mathbb{R}$. For details we refer the reader to [LeVi, Ch. II].

The following function is the solution of the generalization of the Teichmüller problem to $\mathbb{R}^n$. For $x \in \mathbb{R}^n \setminus \{0, e_1\}$, $n \geq 2$, define

$$p(x) = \inf_{E,F} \mathsf{M}(\Delta(E,F)), \tag{2.23}$$

where the infimum is taken over all pairs of continua E and F in $\overline{\mathbb{R}}^n$ with $0, e_1 \in E$, $x, \infty \in F$. Spherical symmetrization yields the inequality [Vu4, 8.11.],

$$p(x) \geq \max\{\tau_n(|x|), \tau_n(|x - e_1|)\} \tag{2.24}$$

for all $x \in \mathbb{R}^n \setminus \{0, e_1\}$ with equality for $x = se_1$, where $s < 0$ or $s > 1$. For $n = 2$ (2.24) was proved by Teichmüller in [Te]. The following upper bounds were given in [Vu4, 8.12]. For $x \in \mathbb{R}^n \setminus \{0, e_1\}$, $|x - e_1| \leq |x|$,

$$p(x) \leq 2\tau_n(|x - e_1|) \text{ when } |x + e_1| \geq 2, \tag{2.25}$$

$$p(x) \leq 4\tau_n(|x - e_1|) \text{ when } |x| \geq 1, \tag{2.26}$$

$$p(x) \leq 2^{n+1}\tau_n(|x - e_1|). \tag{2.27}$$

Betsakos has proved in [Bet] that (2.26) in fact holds for all $x \in \mathbb{R}^n \setminus \{e_1\}$ with $|x - e_1| \leq |x|$. Vuorinen proved in [Vu5, 1.5.] that

$$p(x) \leq \tau_n\left(\frac{|x| + |x - e_1| - 1}{2}\right) \quad \text{for } x \in \mathbb{R}^n \setminus [0, e_1]. \tag{2.28}$$

2.29. **Modulus estimates.** It is not obvious from the definition how $\mathsf{M}(\Delta(E,F))$, for nonempty $E, F \subset \overline{\mathbb{R}}^n$, depends on the geometric setup and the structure of the sets E and F. The following lemma shows that $\mathsf{M}(\Delta(E,F))$ and

$$\frac{\min\{d(E), d(F)\}}{d(E,F)}$$

are simultaneously small or large, provided that E and F are connected. Thus, if E and F are connected, $\mathsf{M}(\Delta(E,F))$ is small if E or F has a small diameter or if these sets are far from each other and in the opposite case $\mathsf{M}(\Delta(E,F))$ is large. It is useful to notice how in this way, in the case of a ring, for instance, one obtains information about the geometry of the ring if its capacity is known. In [Vu4, Ch. 7] one can find earlier results of this type, many of which are based on Gehring's work [Ge1] concerning symmetrization.

2.30. **Lemma.** *Given $n \geq 2$, there exist homeomorphisms $h_j : [0, \infty) \to [0, \infty)$, $j = 1, 2$, with the following property. If E and F are the components of the complement of a nondegenerate ring domain in $\overline{\mathbb{R}}^n$, then*

$$h_1(T) \leq \mathsf{M}(\Delta(E,F)) \leq h_2(T),$$

where

$$T = \frac{\min\{d(E), d(F)\}}{d(E,F)}.$$

Proof. Let

$$g_1(t) = c_n \log(1+t) \quad \text{and} \quad g_2(t) = \tau_n(4/t^2 + 4/t).$$

It was shown in [Vu4, 7.38.] that

$$\mathsf{M}(\Delta(E,F)) \geq g_2(T) \geq g_1(T),$$

so either of the homeomorphisms g_1 and g_2 may be chosen as h_1.

It follows by a straightforward computation from the proof of [Vu3, 2.39.], that we may choose

$$h_2(t) = \begin{cases} \alpha\omega_{n-1}(-\log 2t)^{1-n} & \text{if } t \leq 1/4, \\ \beta 6^n \omega_{n-1} t^n / n & \text{if } t > 1/4, \end{cases}$$

where $\alpha = \max\{1, \gamma\}$, $\beta = \max\{1, 1/\gamma\}$, and $\gamma = (3/2)^n (\log 2)^{n-1}/n$. □

Next we set out to improve the functions h_1 and h_2.

2.31. **Lemma.** *For $r > 1$ let $\Gamma_r = \Delta(\overline{B}^n, S^{n-1}(r))$. Then the following holds.*
(1) $\gamma_n(r) \leq \mathsf{M}(\Gamma_r) = \omega_{n-1}(\log r)^{1-n}$;
(2) $\lim_{r\to 1+} \gamma_n(r)/\mathsf{M}(\Gamma_r) = 0$;
(3) *For each $r_0 > 1$ there exists such a constant $c_1 > 1$, depending on n and r_0, that for all $r > r_0$,*

$$\mathsf{M}(\Gamma_r) < c_1\gamma_n(r) < 2^{n-1}c_1\tau_n(r-1)$$

and $c_1 \to 1$ when $r_0 \to \infty$. In particular,

$$\lim_{r\to\infty} \mathsf{M}(\Gamma_r)/\gamma_n(r) = 1.$$

Proof. (1) The right hand side follows from Lemma 2.6 and the left hand side from the fact that $\Gamma_r < \Delta(\overline{B}^n, [re_1, \infty))$.

(2) By (2.14) and (3.6), we have

$$\begin{aligned} \frac{\gamma_n(r)}{\omega_{n-1}(\log r)^{1-n}} &< \frac{2^{n-1}c_n \log(4(r+1)/(r-1))}{\omega_{n-1}(\log r)^{1-n}} \\ &\leq \frac{2^{n-1}c_n}{\omega_{n-1}} \log\left(4\frac{r+1}{r-1}\right)(r-1)^{n-1} \to 0, \end{aligned}$$

as $r \to 1+$.

(3) For the first inequality, we again apply (2.14) to obtain

$$\begin{aligned} \frac{\mathsf{M}(\Gamma_r)}{\gamma_n(r)} &\leq \frac{\omega_{n-1}(\log r)^{1-n}}{\omega_{n-1}(\log \lambda_n r)^{1-n}} = \left(\frac{\log r + \log\lambda_n}{\log r}\right)^{n-1} \\ &< \left(1 + \frac{\log\lambda_n}{\log r_0}\right)^{n-1} = c_1 \end{aligned}$$

for all $r > r_0$. The second inequality follows from the inequality

$$\gamma_n(r) = 2^{n-1}\tau_n(r^2-1) = 2^{n-1}\tau_n((r-1)(r+1)) < 2^{n-1}\tau_n(r-1),$$

where we used (2.13). Clearly $c_1 \to 1$ as $r_0 \to \infty$. Hence $\limsup_{r\to\infty} \mathsf{M}(\Gamma_r)/\gamma_n(r) \leq 1$ and this together with (1) implies the last claim. □

2.32. **Lemma.** *Let $E, F \subset \mathbb{R}^n$ be continua with $0 < d(E) \le d(F)$. Then the following inequalities hold.*

(1) *If $d(E,F) > d(E)$ then*

$$\mathsf{M}(\Delta(E,F)) \le \omega_{n-1}\left(\log\frac{d(E,F)}{d(E)}\right)^{1-n}.$$

(2) $\mathsf{M}(\Delta(E,F)) \ge \tau_n\left(4\left(\frac{d(E,F)}{d(E)}\right)^2 + 4\frac{d(E,F)}{d(E)}\right) \ge 2^{1-n}\tau_n\left(\frac{d(E,F)}{d(E)}\right).$

(3) *If $d(E,F) > b\,d(E)$ with $b > 1$, then*

$$\mathsf{M}(\Delta(E,F)) \le c_3\,\tau_n\left(\frac{d(E,F)}{d(E)}\right).$$

Here the constant c_3 depends only on n and b.

Proof. Let $x \in E$ and $A = B^n(x, d(E,F)) \setminus \overline{B^n(x, d(E))}$. Then $A \cap E = \emptyset = A \cap F$ and (1) follows from Lemma 2.6. The inequalities in (2) are just a restatement of [Vu4, 7.38] and [Vu4, 7.39].

It remains to prove (3). From Lemma 2.31 (3) it follows that

$$\mathsf{M}(\Delta(E,F)) \le 2^{n-1}\left(1+\frac{\log\lambda_n}{\log b}\right)^{n-1}\tau_n\left(\frac{d(E,F)}{d(E)} - 1\right). \tag{2.33}$$

Next, for $r_0 > 1$ and $r > r_0$ we have, by (2.13), (2.14), and (3.6) that

$$\begin{aligned}\frac{\tau_n(r-1)}{\tau_n(r)} &= \frac{2^{1-n}\gamma_n(\sqrt{r})}{2^{1-n}\gamma_n(\sqrt{1+r})} \le \frac{\omega_{n-1}(\log\sqrt{r})^{1-n}}{\omega_{n-1}(\log\lambda_n\sqrt{1+r})^{1-n}}\\ &= \left(\frac{\log\lambda_n^2 + \log(1+r)}{\log r}\right)^{n-1} \le \left(1+\frac{\log(2\lambda_n^2)}{\log r_0}\right)^{n-1}\\ &= c_2(n, r_0),\end{aligned}$$

where in the last inequality we used the fact that $1 + r \le 2r$. Since $d(E,F)/d(E) > b$, we get from (2.33) that

$$\mathsf{M}(\Delta(E,F)) \le 2^{n-1}\left(1+\frac{\log\lambda_n}{\log b}\right)^{n-1} c_2(n,b)\,\tau_n\left(\frac{d(E,F)}{d(E)}\right).$$

In conclusion, (3) holds with

$$c_3 = 2^{n-1}\left(1+\frac{\log\lambda_n}{\log b}\right)^{n-1}\left(1+\frac{\log(2\lambda_n^2)}{\log b}\right)^{n-1}. \qquad \square$$

2.34. **Corollary.** *Let $E, F \subset \mathbb{R}^n$ be continua with $d(E,F) > b\min\{d(E), d(F)\} > 0$, where $b > 1$. Then*

$$2^{1-n}\tau_n(1/T) \le \tau_n(4/T^2 + 4/T) \le \mathsf{M}(\Delta(E,F)) \le c_3\tau_n(1/T),$$

where $T = \min\{d(E), d(F)\}/d(E,F)$ and c_3 is a constant depending only on n and b. $\square$

2.35. **Remark.** As one can see from the proof of Lemma 2.32, c_3 is a strictly decreasing function of b and $c_3 \to \infty$ as $b \to 1+$ and hence for each $b_0 > 1$,

$$\sup\{c_3(b) \mid b > b_0\} \le c_3(b_0) < \infty. \qquad \square$$

3. The case $n = 2$

In the plane many of the relations presented in Section 2 can be refined and completed.

3.1. **The Teichmüller capacity τ_2.** Using [LeVi, Th 1.1. and Ch 2.1.], the Teichmüller capacity τ_2 can be expressed for $t > 0$ as

$$\tau_2(t) = \frac{\pi}{\mu(1/\sqrt{1+t})}, \tag{3.2}$$

where, for $0 < r < 1$,

$$\mu(r) = \frac{\pi}{2}\frac{\mathcal{K}(\sqrt{1-r^2})}{\mathcal{K}(r)} \tag{3.3}$$

and

$$\mathcal{K}(r) = \int_0^1 \frac{dx}{\sqrt{(1-x^2)(1-r^2x^2)}}, \tag{3.4}$$

which is called *a complete elliptic integral of the first kind.* These functions have been studied systematically in [AVV] where for instance many inequalities are proved for $\mu(r)$. By [LeVi, Ch 2.2.], the function μ satisfies the functional identities

$$\begin{aligned} \mu(r)\,\mu(\sqrt{1-r^2}) &= \frac{1}{4}\pi^2, \\ \mu(r)\,\mu\left(\frac{1-r}{1+r}\right) &= \frac{1}{2}\pi^2, \\ \mu(r) &= 2\,\mu\left(\frac{2\sqrt{r}}{1+r}\right). \end{aligned} \tag{3.5}$$

As for estimates for the function μ, the inequalities

$$\log\frac{1}{r} < \log\frac{1+3\sqrt{1-r^2}}{r} < \mu(r) < \log\frac{2(1+\sqrt{1-r^2})}{r} < \log\frac{4}{r}, \tag{3.6}$$

follow from [LeVi, p. 62] and [AVV, 5.68 (16)].

The Teichmüller capacity function satisfies the duplication formula [AVV, 5.19. (5)]

$$\tau_2(t) = 2\tau_2((\sqrt{t}+\sqrt{t+1})^4 - 1) \tag{3.7}$$

for $t > 0$.

The values of $\mu(t)$ and $\tau_2(t)$ can be computed for certain values of t. For instance, by the first equality in (3.5) we have $\mu(1/\sqrt{2})^2 = \pi^2/4$ and therefore $\mu(1/\sqrt{2}) = \pi/2$. Then other values of $\mu(t)$

can be computed using the *Landen transformations* as explained in [AVV, Chapter 5]. By (3.2) we get $\tau_2(1) = \pi/\mu(1/\sqrt{2}) = 2$. On the other hand, by (3.7) we get $\tau_2(1) = 2\tau_2((1+\sqrt{2})^4 - 1) = 2\tau_2(16+12\sqrt{2})$, so that $\tau_2(16 + 12\sqrt{2}) = 1$.

Next we recall the definition of a function well known and employed in the study of the distortion of plane quasiconformal mappings, see [LeVi, p. 81]. This function is defined for $K \geq 1$ by

$$\lambda(K) = \left(\frac{\mu^{-1}(\pi/(2K))}{\mu^{-1}(\pi K/2)}\right)^2. \tag{3.8}$$

From [AVV, (10.3) and 10.24] it follows that

$$\frac{1}{2}\tau_2(t) \geq \tau_2(\lambda(2)\max\{t^2, t^{1/2}\}), \quad t \in (0, \infty). \tag{3.9}$$

Certain values of $\lambda(K)$ can be expressed in terms of so called *singular values*. For each $p \in \mathbb{N} \setminus \{0\}$, the number $k_p \in (0, 1)$ satisfying

$$\mu(k_p) = \frac{\pi}{2}\frac{\mathcal{K}(\sqrt{1-k_p^2})}{\mathcal{K}(k_p)} = \frac{\pi}{2}\sqrt{p}$$

is called the *pth singular value* of the complete elliptic integral in (3.4) (see [BB, p. 139, 296]). By [AVV, (10.45)] we have that

$$\lambda(\sqrt{p}) = \frac{1-k_p^2}{k_p^2}, \quad p = 1, 2, \ldots.$$

The numbers k_p, $p = 1, \ldots, 9$ are given in [BB, p. 139]. For instance, $k_4 = 3 - 2\sqrt{2}$ and hence we may compute

$$\lambda(2) = \frac{1-(3-2\sqrt{2})^2}{(3-2\sqrt{2})^2} = 16 + 12\sqrt{2}. \tag{3.10}$$

3.11. **The function p.** There is an explicit formula, involving the elliptic integral $\mathcal{K}$, for $p(z)$ when $z \in \mathbb{C} \setminus \{0, 1\}$ (see [SV, (3.3) and 3.5.]). This formula also enables one to compute the values of $p(z)$ by a computer (see [AVV, Remark 15.37.]). The algorithm has been implemented in the program package included in [AVV].

The lower bound (2.24) and the upper bound (2.28) have been improved in [SV, Theorem 1.8.] and [SV, Theorem 1.6.], respectively. Furthermore, Betsakos and Vuorinen have given in [BV, Theorem 1.1] the improvement

$$p(x) \leq \tau_2(2)\tau_2(\min\{|x|, |x - e_1|\}) \tag{3.12}$$

for all $x \in \mathbb{R}^2 \setminus \{0, e_1\}$, where equality holds if and only if $x = e_1/2$. Note that $\tau_2(2) \approx 1.71$.

By [SV, 1.7.] we have the following duplication formula for the function p. Namely, for $z \in \mathbb{C} \setminus \{0, 1\}$ with $\operatorname{Re} z \geq 1/2$ and $\operatorname{Im} z \geq 0$,

$$p(z) = 2p(w^4), \quad w = \sqrt{z} + \sqrt{z-1}, \tag{3.13}$$

where the branches of the square root are chosen so that $0 \leq \arg\sqrt{z} \leq \pi/2$ and $0 \leq \arg\sqrt{z} - 1 \leq \pi$.

Next we show how the functional identity (3.13) combined with the inequality (2.28) leads to a sharper inequality than (2.28) by itself. This procedure succeeds because, roughly speaking, the inequality (2.28) becomes more accurate when $|z|$ increases. However, in this exposition these claims are to be considered only as heuristic principles since, at the time being, they are based only on numerical observations and we do not attempt to prove them here. In [BV], a similar idea based on this duplication formula was employed to improve the lower bound (2.24).

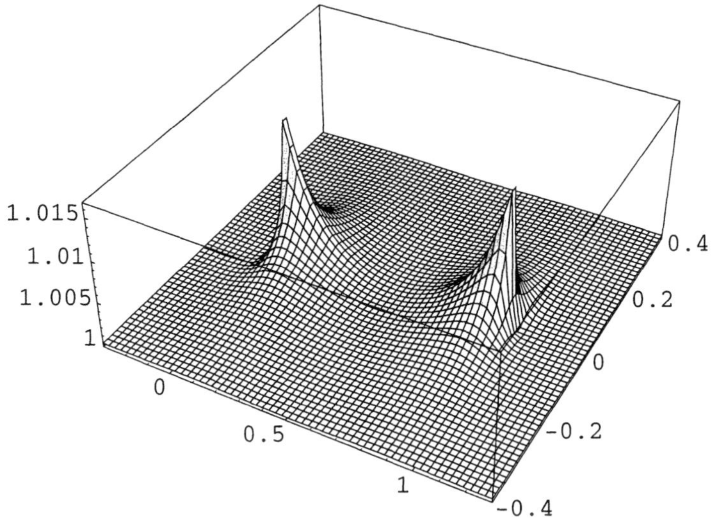

Figure 1: The value of the upper bound in (3.14) divided by the computational value of $p(z)$.

Fix $z \in \mathbb{C}$ with $\operatorname{Re} z \geq 1/2$ and $\operatorname{Im} z \geq 0$ and let $w = \sqrt{z} + \sqrt{z-1}$. By [BV, (2.8) and (2.9)],

$$|w|^4 = A^4 \quad \text{and} \quad |w^4 - 1| = 4tA^2,$$

where $A^2 = |z| + |z-1| + \sqrt{(|z|+|z-1|)^2 - 1}$ and $t = \sqrt{|z||z-1|}$. Then an elementary calculation shows that

$$\frac{|w|^4 + |w^4 - 1| - 1}{2} = u^2A^2 - 1,$$

where $u = \sqrt{|z|} + \sqrt{|z-1|}$. It is easy to see that the mapping $z \mapsto (\sqrt{z} + \sqrt{z-1})^4$ does not map any point in $\mathbb{C} \setminus \{0,1\}$ to 0 or 1. Hence, by [SV, Theorem 1.7.], (2.28), (2.13), and [AVV, 5.19.(2)],

$$\begin{aligned} p(z) &= 2p(w^4) \leq 2\tau_2\left(\frac{|w|^4 + |w^4-1| - 1}{2}\right) \\ &= 2\tau_2(u^2A^2 - 1) = \gamma_2(uA) \qquad (3.14) \\ &= \tau_2\left(\frac{(uA-1)^2}{4uA}\right). \end{aligned}$$

The above derivation of formula (3.14) yields a new proof of [SV, Theorem 1.6.]. The accuracy of the inequality thus obtained is illustrated in Figure 1. As this figure shows, the ratio of the upper bound in (3.14) and the computational value of $p(z)$ is quite close to one and hence the upper bound is very accurate.

3.15. **The modulus.** We first recall a lemma which is due to Gehring.

3.16. **Lemma.**[LeVu, 2.7.] *Suppose that E and F are disjoint compact sets in $\mathbb{R}^2$ and that f is a homeomorphism defined on $E \cup F$ with the following properties:*

$$|f(x) - f(y)| \geq |x - y|, \text{ if } x, y \in E \text{ or if } x, y \in F.$$

$$|f(x) - f(z)| \leq |x - z|, \text{ if } x \in E, z \in F.$$

Let $\Gamma = \Delta(E, F)$. Then $\mathsf{M}(f\Gamma) \geq \mathsf{M}(\Gamma)$. □

It would be interesting to know whether the above lemma holds more generally in $\mathbb{R}^n$, $n \geq 2$. Using this lemma, Lehtinen and Vuorinen proved the following result which provides a useful upper bound and another rare example of a curve family whose modulus can be computed.

3.17. **Lemma.**[LeVu, 2.8.] *Let $x, y \in B^2 \setminus \{0\}$, $|x| \leq |y|$ and $\Gamma = \Delta([0,x],[y,y/|y|];B^2)$, $\Gamma_0 = \Delta([0,x],[|y|x/|x|,x/|x|];B^2)$. Then*

$$\mathsf{M}(\Gamma) \leq \mathsf{M}(\Gamma_0) = \tau_2\left(\frac{(|y|-|x|)(1-|x||y|)}{|x|(1-|y|)^2}\right)$$

with equality if $y = tx$, $t > 1$. □

4. Capacity estimates

When A is a subset of $\mathbb{R}^n$ and $t > 0$, denote by $A + B^n(t)$ the set $\cup_{x \in A} B^n(x,t) = \{z \in \mathbb{R}^n \,|\, d(z,A) < t\}$, which is called the *t-neighborhood* of A. In this section we shall study the capacity

$$\text{(4.1)} \qquad \text{cap}(E + B^n(t), E) = \mathsf{M}(\Delta(\partial(E + B^n(t)), E))$$

as a function of t for compact subsets E of $\mathbb{R}^n$. Later on we shall make additional assumptions on the set E. We start by recalling a result of Väisälä.

4.2. **Lemma.**[Vä3, Lemma 3] *If $E \subset \mathbb{R}^n$ is compact with $\text{cap}E > 0$, then*

$$\text{(4.3)} \qquad \text{cap}(E + B^n(t), E) \to \infty \quad \text{as} \quad t \to 0. \qquad \Box$$

Neither this lemma nor its proof provide any quantitative bounds for the rate of convergence in (4.3). An upper bound for the growth is given by the following lemma.

4.4. **Lemma.**[Vu4, 6.27.] *Let E be a compact set in $B^n(R)$, $R > 0$. Then*

$$\operatorname{cap}(E + B^n(t), E) \leq a(t) \operatorname{cap}(E + B^n(1), E) \tag{4.5}$$

for $t > 0$, where $a(t) = a(1)$ for $t \geq 1$ and $a(t) \leq a_1 t^{-n}$ for $t \in (0,1)$, and a_1 depends only on n and R. □

Vuorinen has also shown in [Vu4, 6.28] that for every $t \in (0,1)$ there exists a compact set $E = E_t \subset [0,1]^n$ satisfying

$$\operatorname{cap}(E + B^n(t), E) \geq \frac{1}{4} t^{-n}.$$

As the next theorem shows, there exist compact sets E in $\mathbb{R}^n$ with positive capacity and with an arbitrarily slow rate of convergence in (4.3).

4.6. **Theorem.** *Let $h : (0,\infty) \to (0,\infty)$ be a decreasing homeomorphism which satisfies $h(t) \to \infty$ as $t \to 0+$. Then there exists a compact set $E \subset \mathbb{R}^n$ with $\operatorname{cap} E > 0$ satisfying*

$$\operatorname{cap}(E + B^n(t), E) < h(t)$$

for all $t \in (0,1)$.

Proof. Let $r_0 = 1$ and define recursively

$$r_{k+1} = \frac{1}{10}(u_k - u_{k+1}), \tag{4.7}$$

where, for $k = 0, 1, 2, \ldots$,

$$u_k = \exp\left(\left(\frac{h((4/3)r_k)}{2\omega_{n-1}}\right)^{1/(1-n)}\right). \tag{4.8}$$

For $r > 0$, let

$$\varphi(r) = r + \frac{1}{10}\exp\left(\left(\frac{h((4/3)r)}{2\omega_{n-1}}\right)^{1/(1-n)}\right).$$

Since h is a decreasing homeomorphism, it follows that $\varphi : (0,\infty) \to (1/10,\infty)$ is an increasing homeomorphism. Given $r_k > 0$, we have that

$$\varphi(0+) = \frac{1}{10} < \frac{1}{10}u_k,$$

$$\varphi(r_k) = r_k + \frac{1}{10}u_k > \frac{1}{10}u_k.$$

Therefore there exists a number $r_{k+1} \in (0, r_k)$ such that

$$\varphi(r_{k+1}) = \frac{1}{10}u_k. \tag{4.9}$$

Hence (4.7) indeed gives a well defined and strictly decreasing positive sequence $(r_k)_{k=0}^{\infty}$. It follows that there exists the limit $\lim_{k\to\infty} r_k = \delta \geq 0$. If we had $\delta > 0$, then (4.9) would imply by continuity that $\varphi(\delta) = (1/10)\exp((h((4/3)\delta)/(2\omega_{n-1}))^{1/(1-n)})$, which is a contradiction with the definition of φ. Hence $r_k \to 0$ as $k \to \infty$. Thus $(u_k)_{k=0}^{\infty}$ is a strictly decreasing sequence with $u_k > 1$ for all k and $u_k \to 1$ as $k \to \infty$. From (4.7) we see that $u_k + 10r_k = u_{k-1}$, which implies that $u_k + r_k < u_{k-1}$ for $k = 1, 2, \ldots$.

For each $i \in \mathbb{N}$, let C_i^m, $m \in \mathbb{N}$, consist of all points in $R_i = \overline{B^n(u_i)} \setminus B^n(u_{i+1} + r_{i+1})$ with each coordinate of the form $a/2^m$, $a \in \mathbb{Z}$. Since R_i is a compact subset of $\mathbb{R}^n$, C_i^m is a finite set for each $m \in \mathbb{N}$. We choose $m_i \in \mathbb{N}$ to be large enough for the following to hold. For each $x \in C_i^{m_i}$,

$$S^{n-1}(x, (4/3)r_{i+1}) \cap \complement \left(\bigcup_{z \in C_i^{m_i} \setminus \{x\}} \overline{B^n(z, (4/3)r_{i+1})} \right) \subset \complement \left(\overline{B^n(u_i)} \setminus B^n(u_{i+1}) \right). \tag{4.10}$$

The condition (4.10) implies that if $t \geq (4/3)r_{i+1} > r_{i+1}$, then

$$\partial(C_i^{m_i} + B^n(t)) \cap (\overline{B^n(u_i)} \setminus B^n(u_{i+1})) = \emptyset.$$

We denote $C_i = C_i^{m_i}$ and define

$$E = \overline{B^n} \cup \bigcup_{i=1}^{\infty} C_i.$$

The set $\complement E$ is clearly an open set and hence E is closed. Since E is also bounded, it is compact. The set E has positive capacity, since $\operatorname{cap}\overline{B^n} > 0$. Let $t \in (0, 1)$. We find an index $k \in \mathbb{N}$ such that $t \in [(4/3)r_{k+1}, (4/3)r_k)$. Now, by the construction of the set E, there exist no components of $\partial(E + B^n(t))$ in $B^n(u_k)$. Furthermore, the set $\cup_{i=1}^{\infty} C_i$ is a countable set consisting of discrete points and hence has no effect in computing $\operatorname{cap}(E + B^n(t), E)$. Since h is a strictly decreasing function, we get that

$$\begin{aligned} \operatorname{cap}(E + B^n(t), E) &\leq \operatorname{cap}(B^n(u_k), \overline{B^n}) = \omega_{n-1}(\log u_k)^{1-n} \\ &= h((4/3)r_k)/2 \leq h(t)/2 \\ &< h(t). \end{aligned}$$

Hence the set E satisfies the requirements of the theorem. □

From Theorem 5.1 below one easily obtains the following corollary.

4.11. **Lemma.** *Let $\delta > 0$ and let E be a compact set in $\mathbb{R}^n$ with $0 \in E$. Assume that for all $t \in (0, 1)$, $\operatorname{cap}(0, E, t) \geq \delta$ and that there exists a continuum $F_t \subset \mathbb{R}^n \setminus (E + B^n(t))$ with $F_t \cap S^{n-1}(t) \neq \emptyset \neq F_t \cap S^{n-1}$. Then, for all $t \in (0, 1)$,*

$$\operatorname{cap}(E + B^n(t), E) \geq d^* \log \frac{1}{t},$$

where d^* is a positive constant depending only on n and δ. □

4.12. **Remarks.** (1) If $E \subset H^n \cup \{0\}$, then the continua F_t with the properties of Lemma 4.11 can be produced from a single continuum. Namely, one can choose F to be the line segment $[0, -e_n]$ and let $F_t = F \setminus (E + B^n(t))$ for $t \in (0,1)$.

(2) Let E be the set constructed in the proof of Theorem 4.6. Then we have that

$$\begin{aligned} \operatorname{cap}(0, E + e_n, t) &\geq \operatorname{cap} R(\mathbb{R}^n \setminus B^n(2t), [0, te_n]) \\ &= \operatorname{cap} R(\mathbb{R}^n \setminus B^n, [0, e_1/2]) \\ &= R^b_{G,n}(1/2) = \gamma_n(2) > 0, \end{aligned}$$

where the first equality was obtained by applying a stretching and a rotation. Note that the existence of the continuum F in Lemma 4.11 fails to hold. □

The goal of the remainder of this section is to obtain lower bounds and to improve the upper bound of the type t^{-n} for the capacity (4.1) when the set E satisfies some additional assumptions. To this end, we first recall some definitions.

4.13. **Hausdorff content, measure, and dimension.** Let E be a set in $\mathbb{R}^n$ and $\beta > 0$. Then the *β-dimensional Hausdorff content* of E is

$$\Lambda^\beta(E) = \inf\left\{\sum_{i=1}^{\infty} r_i^\beta\right\},$$

where the infimum is taken over all coverings of E with countably many Euclidean balls of radii r_i. Set

$$H^\beta_\delta(E) = \inf\left\{\sum_{i=1}^{\infty} d(U_i)^\beta\right\},$$

where the infimum is taken over all countable coverings of E by such sets U_i that $d(U_i) < \delta$. Then the *β-dimensional Hausdorff measure* of E is

$$H^\beta(E) = \lim_{\delta \to 0} H^\beta_\delta(E).$$

The *Hausdorff dimension* of E is defined as

$$\dim_H(E) = \inf\{\beta \mid H^\beta(E) < \infty\}.$$

It is clear that $\Lambda^\beta(E) \leq H^\beta(E)$ for all $E \subset \mathbb{R}^n$ and $\beta > 0$ and easy to prove that $\Lambda^\beta(E) = 0$ if and only if $H^\beta(E) = 0$, see [Re2, p. 114]. As will be pointed out in Remark 4.20, a reverse inequality also holds for Ahlfors regular sets, which are defined in 4.14 below.

Let $d \in (0,n)$, $d' \in (d,n)$. By [Beal, Theorem 7], there exists a general Cantor set $E \subset B^n(0,1)$ with $\dim_H(E) = d'$. Since $E \subset B^n(0,1)$, we have $\Lambda^d(E) \leq 1$, and since $H^d(E) = \infty$ and H^d and Λ^d are simultaneously zero, we also have $\Lambda^d(E) > 0$. Hence E is a set satisfying

$$0 < \Lambda^d(E) < H^d(E) = \infty.$$

Sets of capacity zero have Hausdorff dimension zero by Lemma 2.9 but the converse does not hold. Namely, from [Wa, 4.4.] it follows that there exists a compact set $E \subset \mathbb{R}^n$ with cap$E > 0$ but $H^\beta(E) = 0$ for all $\beta > 0$.

4.14. **Ahlfors regularity.** [DS] A subset E of $\mathbb{R}^n$ is called *(Ahlfors) regular with dimension* d if it is closed in $\mathbb{R}^n$ and if there exists a constant $C_0 > 0$ such that

$$C_0^{-1} r^d \le H^d(E \cap B^n(x,r)) \le C_0 r^d$$

for all $x \in E$ and $r \in (0, d(E))$.

The Ahlfors regularity of a set means that, in a sense, the mass of the set is distributed evenly. The significance of the concept of Ahlfors regularity is that it describes a notion of dimension. For instance d-dimensional hyperplanes in $\mathbb{R}^n$, $d \le n$, are Ahlfors regular with dimension d.

4.15. **Uniform perfectness.** Let $\alpha > 0$ and assume that $E \subset \overline{\mathbb{R}}^n$ is a closed set containing at least two points. Then E is *α-uniformly perfect* if there is no ring domain $D \subset \overline{\mathbb{R}}^n$ separating E with mod$D > \alpha$. E is *uniformly perfect* if it is α-uniformly perfect for some $\alpha > 0$.

Uniform perfectness is a useful tool in many topics of geometric function theory. See [Su] for a survey of this topic. For our purposes uniformly perfect sets provide a class of sets that "interpolates" between connected sets and sets of positive capacity.

In their pioneering paper, Järvi and Vuorinen presented several characterizations of uniform perfectness. We will repeatedly invoke the ones contained in the following theorem.

4.16. **Theorem.**[JV, 4.1.] *Let $E \subset \overline{\mathbb{R}}^n$ be a closed set containing at least two points. Then the following properties are equivalent:*

(1) *E is α-uniformly perfect for some $\alpha > 0$;*

(2) *there exist positive constants β and C_1 such that $\Lambda^\beta(\overline{B^n(x,r)} \cap E) \ge C_1 r^\beta$ for $x \in E \cap \mathbb{R}^n$ and $r \in (0, d(E))$;*

(3) *there is a positive constant C_2 such that* cap$(x, E, r) \ge C_2$ *for $x \in E \cap \mathbb{R}^n$ and $r \in (0, d(E))$.*

The constants α, β, C_1 and C_2 depend only on n and each other. □

The conditions in 4.14 and 4.15 are related in the following way.

4.17. **Theorem.** *If E is an Ahlfors regular subset of $\mathbb{R}^n$, then $\overline{E}$ is uniformly perfect.*

Proof. Let E be Ahlfors regular with dimension d and parameter $C_0 > 0$. By definition, $\overline{E}$ is closed (in $\overline{\mathbb{R}}^n$) as is required for uniform perfectness. By Theorem 4.16 it is enough to find such a constant $C_1 > 0$ that

$$\Lambda^d(\overline{B^n(x,r)} \cap \overline{E}) = \Lambda^d(\overline{B^n(x,r)} \cap E) \ge C_1 r^d$$

for all $x \in \overline{E} \cap \mathbb{R}^n = E$ and $r \in (0, d(E))$. Let x and r be as said and let $(B^n(x_i, r_i))_{i=1}^\infty$ be a covering of $\overline{B^n(x,r)} \cap E$. We may assume that $B^n(x_i, r_i) \cap E \ne \emptyset$ for all $i \in \mathbb{N} \setminus \{0\}$. For all $i \in \mathbb{N} \setminus \{0\}$, if $x_i \notin E$, choose a point $x_i' \in B^n(x_i,r_i)\cap E$ and put $x_i' = x_i$ if $x_i \in E$. Then $B^n(x_i,r_i)\cap E \subset B^n(x_i',2r_i)\cap E$ for all i, and the center points x_i' are in E. Since H^d is a measure and E is Ahlfors regular, we obtain

$$\begin{aligned} C_0^{-1}r^d &\le H^d(E\cap B^n(x,r)) \\ &\le H^d(E\cap \overline{B^n(x,r)}) \\ &= H^d(\cup_{i=1}^{\infty}(B^n(x_i,r_i)\cap(E\cap\overline{B^n(x,r)}))) \\ &\le H^d(\cup_{i=1}^{\infty}(B^n(x_i,r_i)\cap E)) \\ &\le \sum_{i=1}^{\infty} H^d(B^n(x_i,r_i)\cap E) \\ &\le \sum_{i=1}^{\infty} H^d(B^n(x_i',2r_i)\cap E) \\ &\le C_0 2^d \sum_{i=1}^{\infty} r_i^d. \end{aligned} \tag{4.18}$$

We take the infimum over all countable coverings $(B^n(x_i,r_i))_{i=1}^{\infty}$ of $\overline{B^n(x,r)}\cap E$ and get

$$\Lambda^d(\overline{B^n(x,r)}\cap E) \ge C_1 r^d, \tag{4.19}$$

where $C_1 = C_0^{-2}2^{-d} > 0$. □

4.20. **Remark.** Let E be Ahlfors regular with dimension d and parameter $C_0 > 0$. Replacing $E\cap\overline{B^n(x,r)}$ by E in the proof of Theorem 4.17, we see from the inequalities (4.18), that

$$H^d(E) \le C_0 2^d \Lambda^d(E). \quad \square$$

The proof of Theorem 4.17 is based on Theorem 4.16, which gives no explicit connection between its parameters. Hence we get no formula connecting the parameters of Ahlfors regularity and uniform perfectness. This situation can be improved, as in Theorem 4.21 below, by using a different reasoning which was suggested to the author by Väisälä.

4.21. **Theorem.** *If $A \subset \mathbb{R}^n$ is Ahlfors regular with dimension d and additional parameter $C_0 > 0$, then $\overline{A}$ is δ-uniformly perfect with*

$$\delta = \left(\frac{\tau_n((2C_0^{2/d}+1)^2-1)}{\omega_{n-1}}\right)^{1/(1-n)}.$$

Proof. Let $A \subset \mathbb{R}^n$ be Ahlfors regular with dimension d and parameter $C_0 > 0$. Assume that $\overline{A}$ can be separated by a ring $R = R(E_1,E_2)$ with $\operatorname{mod} R = \alpha$ arbitrarily large. By Lemma 2.30 we have

$$\alpha = \left(\frac{\mathsf{M}(\Delta(E_1,E_2))}{\omega_{n-1}}\right)^{1/(1-n)} \le \left(\frac{\tau_n(4m^2+4m)}{\omega_{n-1}}\right)^{1/(1-n)},$$

where $m = d(E_1, E_2)/\min\{d(E_1), d(E_2)\}$. This is equivalent to

$$4m^2 + 4m - \tau_n^{-1}(\alpha^{1-n}\omega_{n-1}) \geq 0,$$

which is satisfied for

$$m \geq \frac{-1 + \sqrt{1 + \tau_n^{-1}(\alpha^{1-n}\omega_{n-1})}}{2} = c(\alpha, n).$$

If $\alpha > (\tau_n(8)/\omega_{n-1})^{1/(1-n)}$, then $c(\alpha, n) > 1$ and

$$d(E_1, E_2) \geq c(\alpha, n) \min\{d(E_1), d(E_2)\} > \min\{d(E_1), d(E_2)\}.$$

Assume that $d(E_1) \leq d(E_2)$ and choose $x \in E_1$. Then $E_1 \subset \overline{B^n(x, d(E_1))}$ and $B^n(x, d(E_1)) \subset \complement E_2$, and $B^n(x, c(\alpha, n)d(E_1)) \setminus \overline{B^n(x, d(E_1))}$ separates $\overline{A}$. Choose $c_1 \in (1, c(\alpha, n))$. Then

$$\overline{A} \cap \overline{B^n(x, c_1 d(E_1))} = \overline{A} \cap \overline{B^n(x, d(E_1))}$$

and by the Ahlfors regularity of A we get

$$\begin{aligned} C_0^{-1}(c_1 d(E_1))^d &\leq H^d(A \cap \overline{B^n(x, c_1 d(E_1))}) \\ &\leq H^d(\overline{A} \cap \overline{B^n(x, c_1 d(E_1))}) \\ &= H^d(\overline{A} \cap \overline{B^n(x, d(E_1))}) \\ &= H^d(A \cap \overline{B^n(x, d(E_1))}) \\ &\leq C_0 d(E_1)^d. \end{aligned}$$

Hence $c_1^d \leq C_0^2$, which implies $c_1 \leq C_0^{2/d}$. Since c_1 can be chosen arbitrarily close to $c(\alpha, n)$, we must have $c(\alpha, n) \leq C_0^{2/d}$, or equivalently,

$$\alpha \leq \left(\frac{\tau_n((2C_0^{2/d} + 1)^2 - 1)}{\omega_{n-1}} \right)^{1/(1-n)} = \delta.$$

This contradiction concludes the proof. □

4.22. **Example.** Let

$$E = \overline{B^n} \cup [e_1, 2e_1] \subset \mathbb{R}^n.$$

We claim that E is uniformly perfect but not Ahlfors regular with any dimension $d > 1$.

Since E is closed and connected, it is indeed α-uniformly perfect for all $\alpha > 0$.

Let $d > 1$. Then $E \cap B^n(2e_1, 1) = (e_1, 2e_1]$, which set can be covered with the collection

$$\{B^n(1 + i/m, 1/m) \mid i = 1, \ldots, m\}$$

of balls for $m \geq 1$. Hence

$$\Lambda^d((e_1, 2e_1]) \leq \sum_{i=1}^{m} \left(\frac{1}{m}\right)^d = m^{1-d} \to 0$$

as $m \to \infty$, since $d > 1$. Since Λ^d and H^d are simultaneously zero, we have

$$H^d(E \cap B^n(2e_1, 1)) = H^d((e_1, 2e_1]) = 0,$$

which implies that E cannot be Ahlfors regular with dimension d.

Note that E has a subset, namely $\overline{B^n}$, which is Ahlfors regular with dimension n and another subset, the line segment $[e_1, 2e_1]$, which fails to be Ahlfors regular for all dimensions > 1. Thus it would be sufficient to consider only the line segment $[e_1, 2e_1]$. □

For a nonempty bounded subset E of $\mathbb{R}^n$ and $\varepsilon > 0$, denote by $N(E, \varepsilon)$ the smallest number of balls $B^n(x, \varepsilon)$, $x \in E$, required to cover E and by $P(E, \varepsilon)$ the greatest number of disjoint balls $B^n(x, \varepsilon)$ with $x \in E$. This notation is adopted from [Mat, 5.3] with the slight modification that in the definition of $N(E, \varepsilon)$ used here, the centerpoints of the covering balls are in E.

4.23. **Lemma.** *If $E \subset \mathbb{R}^n$ is compact and Ahlfors regular with dimension d, then*

$$N(E, t) \le \frac{C_0 2^d H^d(E)}{t^d}$$

for all $t \in (0, d(E))$, where C_0 is the constant in definition 4.14.

Proof. Let $t \in (0, d(E))$. We choose disjoint balls $B^n(x_i, t/2)$, $x_i \in E$, $i = 1, \ldots, k$, $k = P(E, t/2)$. If there existed a point $x \in E \setminus \cup_{i=1}^k B^n(x_i, t)$, then the balls $B^n(x_1, t/2), \ldots, B^n(x_k, t/2), B^n(x, t/2)$ would be disjoint, which is absurd by the definition of $P(E, t/2)$. Hence the balls $B^n(x_i, t)$ cover E and it follows that $N(E, t) \le P(E, t/2)$.

Since E is Ahlfors regular with dimension d, we have that

$$C_0^{-1} P(E, t/2)(t/2)^d \le \sum_{i=1}^{P(E,t/2)} H^d(E \cap B^n(x_i, t/2)) \le H^d(E).$$

We conclude that

$$N(E, t) \le P(E, t/2) \le \frac{C_0 2^d H^d(E)}{t^d}. \qquad \square$$

4.24. **Theorem.** *Let $E \subset \mathbb{R}^n$ be compact and Ahlfors regular with dimension d. Then*

$$\operatorname{cap}(E + B^n(t), E) \le \frac{c}{t^d}$$

for all $t \in (0, d(E))$, where $c = C_0\, 4^d H^d(E)\, \omega_{n-1} (\log 2)^{1-n}$ and C_0 is the constant in definition 4.14.

Proof. Let $t \in (0, d(E))$. We choose a covering for E by balls $B^n(x_i, t/2)$, $x_i \in E$, $i = 1, \ldots, k$, $k = N(E, t/2)$. We denote

$$\Gamma_t = \Delta(E, \partial(E + B^n(t)));$$

$$\Gamma_i = \Delta(E \cap B^n(x_i, t/2), S^{n-1}(x_i, t));$$

$$\Delta_i = \Delta(\overline{B^n(x_i,t/2)}, S^{n-1}(x_i,t)).$$

Then $\cup_{i=1}^k \Gamma_i < \Gamma_t$ and $\Delta_i < \Gamma_i$ for all i. Hence we get

$$\begin{aligned}\operatorname{cap}(E+B^n(t),E) &= \mathsf{M}(\Gamma_t) \le \mathsf{M}\left(\bigcup_{i=1}^k \Gamma_i\right)\\ &\le \sum_{i=1}^k \mathsf{M}(\Gamma_i) \le N(E,t/2)\,\mathsf{M}(\Delta_i)\\ &= N(E,t/2)\,\omega_{n-1}(\log 2)^{1-n}.\end{aligned}$$

Using the upper bound of Lemma 4.23 for $N(E,t/2)$ we conclude that

$$\operatorname{cap}(E+B^n(t),E) \le \frac{c}{t^d},$$

where $c = C_0\, 4^d H^d(E)\,\omega_{n-1}(\log 2)^{1-n}$. □

4.25. **Lemma.** *Let E be a compact, α-uniformly perfect subset of $\mathbb{R}^n$, where $\alpha > 0$. Then*

$$N(E,t) \ge \frac{C_1 d(E)^\beta}{t^\beta}$$

for all $t > 0$, where C_1 and β are the numbers in Theorem 4.16.

Proof. Let $t > 0$, $x \in E$, $r \in (0, d(E))$. Any covering of E with balls $B^n(y,t)$, $y \in E$, is also a covering of $E \cap \overline{B^n(x,r)}$. Hence

$$N(E,t)t^\beta \ge \Lambda^\beta(E \cap \overline{B^n(x,r)}) \ge C_1 r^\beta$$

by Theorem 4.16. Since this inequality holds for all $r \in (0, d(E))$ and the left hand side is independent of r, we get the desired result. □

4.26. **Porous sets.** A set A in $\mathbb{R}^n$ is *α-porous in $\mathbb{R}^n$*, $0 < \alpha \le 1$, if each closed ball $\overline{B^n(x,r)}$ in $\mathbb{R}^n$ contains a point z such that the open ball $B^n(z,\alpha r)$ does not meet A. A set A in $\mathbb{R}^n$ is called *porous*, if it is α-porous for some α, $0 < \alpha \le 1$.

Porous sets were introduced in [GLM] under the name *thin* sets and their connections with the harmonic measure as well as uniform domains were studied. We note that, as proved in [Sa2], for an α-porous set $A \subset \mathbb{R}^n$, we have $\dim_H(A) \le c(n,\alpha) < n$.

If the set under investigation is porous, then we get the following estimate from below for $\operatorname{cap}(E + B^n(t), E)$. We point out that in general, the number ξ_t in Lemma 4.27 depends on t and may well tend to zero as t tends to zero. As will be shown in the proof of Theorem 4.28, ξ_t has a positive lower bound independent of t for uniformly perfect sets.

4.27. **Lemma.** *Let E be a compact, d-porous subset of $\mathbb{R}^n$ with at least two points, where $0 < d \le 1$. Then, for $t > 0$,*

$$\operatorname{cap}(E + B^n(t), E) \ge N(E, 20t/d)\xi_t,$$

with $\xi_t = \inf \mathsf{M}(\Delta(E \cap \overline{B^n(x,2t/d)}, \overline{B^n(z,t)}; \overline{B^n(x,4t/d)}))$, where the infimum is taken over all $x \in E$ and over all $z \in B^n(x,2t/d)$ such that $B^n(z,2t) \cap E = \emptyset$.

Proof. In what follows we will employ the following notation. If $B = B(x,r)$ is a ball in $\mathbb{R}^n$, then for $a > 0$, $aB = B(x, ar)$.

Let $t > 0$ and denote $t' = 20t/d$, $\mathcal{B} = \{\overline{B^n(x,t'/5)} \mid x \in E\}$. Using [Mat, 2.1.] and the compactness of E, we get a finite sequence $(B_i)_{i=1}^p$, $p \geq 1$, of disjoint balls $B_i = \overline{B^n(x_i, t'/5)}$, $x_i \in E$, such that

$$E \subset \bigcup_{B \in \mathcal{B}} B \subset \bigcup_{i=1}^p 5B_i.$$

Then clearly $p \geq N(E, t')$.

Since E is d-porous, we find in each $\frac{1}{2}B_i$ a point z_i such that $B^n(z_i, dt'/10) \cap E = \emptyset$. Then, for each i, $1 \leq i \leq p$, we have that

$$B^n(z_i, dt'/20) \cap (E + B^n(dt'/20)) = \emptyset$$

and that

$$d(B^n(z_i, dt'/20), S^{n-1}(x_i, t'/5)) \geq t'/5 - t'/10 - dt'/20 = \frac{2-d}{20} t',$$

where $(2-d)/20 > 0$.

Denote

$$\Gamma = \Delta(E, \partial(E + B^n(dt'/20)));$$

$$\Gamma_i = \Delta(E \cap \overline{B^n(x_i, t'/10)}, \partial(E + B^n(dt'/20)); \overline{B^n(x_i, t'/5)});$$

$$\tilde{\Gamma}_i = \Delta(E \cap \overline{B^n(x_i, t'/10)}, \overline{B^n(z_i, dt'/20)}; \overline{B^n(x_i, t'/5)})$$

for $1 \leq i \leq p$, and choose $\delta = \min_{i=1,\ldots,p} \mathsf{M}(\tilde{\Gamma}_i)$. Then clearly $\delta \geq \xi_t$. It is immediate from the definitions of Γ, Γ_i, and $\tilde{\Gamma}_i$, that $\Gamma \supset \cup_{i=1}^p \Gamma_i$, that Γ_i and Γ_j are separate for $i \neq j$, and that $\Gamma < \Gamma_i < \tilde{\Gamma}_i$ for all i, $1 \leq i \leq p$. These facts together with the basic properties of the modulus yield

$$\begin{aligned}
\operatorname{cap}(E + B^n(t), E) &= \operatorname{cap}(E + B^n(dt'/20), E) \\
&= \mathsf{M}(\Gamma) \geq \mathsf{M}\left(\bigcup_{i=1}^p \Gamma_i\right) \geq \sum_{i=1}^p \mathsf{M}(\Gamma_i) \\
&\geq \sum_{i=1}^p \mathsf{M}(\tilde{\Gamma}_i) \geq p\delta \geq N(E, t')\xi_t \\
&= N(E, 20t/d)\xi_t. \qquad \square
\end{aligned}$$

4.28. **Theorem.** *Let E be a compact, α-uniformly perfect and d-porous subset of $\mathbb{R}^n$, where $\alpha > 0$ and $0 < d \leq 1$. Then there exist positive constants β and c', depending only on n, α, and d, such that*

$$\operatorname{cap}(E + B^n(t), E) \geq \frac{c' d(E)^\beta}{t^\beta}$$

for all $t \in (0, d(E))$.

Proof. It follows from Lemmas 4.27 and 4.25, that

$$\begin{aligned}\operatorname{cap}(E+B^n(t),E) &\geq N(E,20t/d)\xi_t\\ &\geq \frac{C_1 d(E)^\beta}{(20t/d)^\beta}\xi_t\\ &= \frac{C_1(d/20)^\beta \xi_t d(E)^\beta}{t^\beta},\end{aligned}$$

where C_1, d, and β depend only on n, α, and d. Hence it suffices to find for ξ_t a positive lower bound which is independent of t. For this fix $x \in E$, $z \in B^n(x, 2t/d)$, $B^n(z,2t) \cap E = \emptyset$. This is possible since E is d-porous. Denote $t' = 20t/d$,

$$\begin{aligned}F_1 &= E \cap \overline{B^n(x,t'/10)},\\ F_2 &= \overline{B^n(z,dt'/20)},\\ F_3 &= F_4 = S^{n-1}(x,t'/5),\\ \Gamma_{hk} &= \Delta(F_h,F_k;\overline{B^n(x,t'/5)}),\ \ h,k\in\{1,2,3,4\},\ h\neq k.\end{aligned}$$

Then, using the comparison principle for the modulus (see [Vu4, 5.35.]), we obtain

$$\mathsf{M}(\Delta(E\cap\overline{B^n(x,2t/d)},\overline{B^n(z,t)};\overline{B^n(x,4t/d)})) = \mathsf{M}(\Gamma_{12})$$

$$\geq 3^{-n}\min\{\mathsf{M}(\Gamma_{13}),\mathsf{M}(\Gamma_{24}),\inf \mathsf{M}(\Delta(|\gamma_{13}|,|\gamma_{24}|;B^n(x,t'/5)))\},$$

where the infimum is taken over all rectifiable curves $\gamma_{13} \in \Gamma_{13}$ and $\gamma_{24} \in \Gamma_{24}$. By Theorem 4.16, there exists a positive constant C_2 such that

$$\mathsf{M}(\Gamma_{13}) = \operatorname{cap}(x,E,t'/10) \geq C_2.$$

Applying the Möbius mappings $\zeta \mapsto \zeta - x$, $\zeta \mapsto (5/t')\zeta$, and $T_{5(z-x)/t'}$ (see [Vu4, 1.34]), we see that

$$\mathsf{M}(\Gamma_{24}) = \mathsf{M}(\Delta(\overline{B^n(0,r)},S^{n-1})),$$

where

$$r = \left|T_{5(z-x)/t'}\left(\frac{5(z-x)}{t'} + \frac{d}{4}\frac{(5/t')(z-x)}{|(5/t')(z-x)|}\right)\right|.$$

Then r can be computed, using [Vu4, 1.41], to be

$$r = (d/4)/(1-(25/t'^2)|z-x|^2 - (5d/4t')|z-x|).$$

Hence

$$\begin{aligned}\mathsf{M}(\Gamma_{24}) &= \omega_{n-1}(\log(4/d - (100/dt'^2)|z-x|^2 - (5/t')|z-x|))^{1-n}\\ &\geq \omega_{n-1}\left(\log\frac{4}{d}\right)^{1-n} > 0.\end{aligned}$$

For each $\gamma_{13} \in \Gamma_{13}$ and $\gamma_{24} \in \Gamma_{24}$, $|\gamma_{13}| \cap S^{n-1}(x,\delta) \neq \emptyset$ and $|\gamma_{24}| \cap S^{n-1}(x,\delta) \neq \emptyset$ for all

$$\begin{aligned} \delta &\in (\max\{t'/10, t'/5 - ((2-d)/20)\, t'\}, t'/5) \\ &= (\max\{t'/10, ((2+d)/20)\, t'\}, t'/5) \\ &= (((2+d)/20)\, t', t'/5). \end{aligned}$$

By Lemma 2.7 we get

$$\begin{aligned} \mathsf{M}(\Delta(|\gamma_{13}|, |\gamma_{24}|; B^n(x,t'/5))) &\geq \mathsf{M}(\Delta(|\gamma_{13}|, |\gamma_{24}|; B^n(x,t'/5) \setminus \overline{B^n(x,((2+d)/20)\, t')})) \\ &\geq c_n \log \frac{t'/5}{((2+d)/20)t'} = c_n \log \frac{4}{2+d} > 0. \end{aligned}$$

We conclude that

$$\mathsf{M}(\Gamma_{12}) \geq C_3 = 3^{-n} \min \left\{ C_2, \omega_{n-1} \left(\log \frac{4}{d} \right)^{1-n}, c_n \log \frac{4}{2+d} \right\} > 0,$$

which is a strictly positive lower bound for $\mathsf{M}(\Gamma_{12})$, independent of x, z, and t. It follows that $\xi_t \geq C_3 > 0$. Hence we may choose $c' = C_1 C_3 (d/20)^\beta$ and the theorem is proved. □

4.29. **Lemma.** *Let $E \subset \mathbb{R}^n$ be Ahlfors regular with dimension d and parameter $C_0 > 0$. Then*

$$N(E,t) \geq \frac{C}{t^d}$$

for all $t > 0$, where $C = d(E)^d C_0^{-2} 2^{-d}$.

Proof. Let $t > 0$, $x \in E$, $r \in (0, d(E))$. Any covering of E with euclidean balls $B^n(y,t)$, $y \in E$, covers also $E \cap \overline{B^n(x,r)}$. Hence

$$N(E,t)t^d \geq \Lambda^d(E \cap \overline{B^n(x,r)}) \geq C_0^{-2} 2^{-d} r^d,$$

where the second inequality follows from (4.19). Letting $r \to d(E)$, we get

$$N(E,t) \geq \frac{d(E)^d C_0^{-2} 2^{-d}}{t^d}. \quad \square$$

4.30. **Theorem.** *Let $E \subset \mathbb{R}^n$ be compact, Ahlfors regular with dimension d and parameter $C_0 > 0$ and a-porous with $0 < a \leq 1$. Then*

$$\operatorname{cap}(E + B^n(t), E) \geq \frac{\tilde{c}}{t^d}$$

for all $t \in (0, d(E))$, where $\tilde{c} = d(E)^d C_0^{-2} 40^{-d} a^d C_3 > 0$ and C_3 is a positive constant depending only on n, a, d, and C_0.

Proof. By Lemmas 4.27 and 4.29 we get

$$\begin{aligned}\operatorname{cap}(E+B^n(t),E) &\geq N(E,20t/a)\xi_t \\ &\geq \frac{d(E)^d C_0^{-2}2^{-d}}{(20t/a)^d}\xi_t \\ &= \frac{d(E)^d C_0^{-2}40^{-d}a^d\xi_t}{t^d}.\end{aligned}$$

Since E is uniformly perfect by Theorem 4.17, we get from the proof of Theorem 4.28 such a positive constant C_3, depending only on n, a, d, and C_0, that $\xi_t \geq C_3 > 0$. Hence $\tilde{c} > 0$ and the proof is complete. □

The following result follows from the work of Mattila and Rickman. Note that here we need not assume porosity.

4.31. **Theorem.** *Let $E \subset \mathbb{R}^n$ be compact and Ahlfors regular with dimension $d > 0$ and parameter $C_0 > 0$. Assume that $E \subset B^n(r)$ with $0 < r < d(E)$. Then*

$$\Lambda^d(E) \leq Dr^d(\operatorname{cap}(B^n(r),E))^p \tag{4.32}$$

for all $p > 2$, where D is a positive constant which depends only on n, p, and d.

Proof. We will apply [MR, Lemma 4.2.] and we start by verifying that the assumptions necessary for this lemma are fulfilled. We choose $N = \overline{B^n(r)}$ and $\mu = \chi_E H^d$, where χ_E denotes the characteristic function of E. It is well known that H^d, and hence also μ, is a Borel measure. Since $E \subset B^n(r)$ and E is Ahlfors regular, we have that

$$\mu(\overline{B^n(r)}) = H^d(E \cap \overline{B^n(r)}) = H^d(E \cap B^n(r)) \in [C_0^{-1}r^d, C_0 r^d],$$

which implies that $0 < \mu(N) < \infty$. Furthermore, μ is h-calibrated, where $h(r) = C_0 r^d$ is a calibration function in the sense of [MR, Section 1]. We check that h satisfies [MR, (1.4)] which is to say that

$$\infty > \int_0^1 \frac{h(r)^{1/(pn)}}{r}dr = C_0^{1/(pn)}\int_0^1 r^{d/(pn)-1}dr.$$

This holds whenever $d/(pn) - 1 > -1$, so all values $p > 0$ are acceptable. Let $\varepsilon > 0$. Since $N = \overline{B^n(r)} \subset \mathbb{R}^n$, we can choose $r_0 = r + \varepsilon$ and the chart map φ_ζ to be a translation in [MR, 3.1].

With the notation γ_h of [MR, Lemma 4.2.] we have

$$\gamma_h(E) = \inf\sum_{i=1}^{\infty} h(r_i) = C_0 \inf\sum_{i=1}^{\infty} r_i^d,$$

where the infimum is taken over all coverings of E with at most a countable number of balls $B^n(x_i, r_i)$, $x_i \in \mathbb{R}^n$, $r_i > 0$. Hence $\gamma_h(E) = C_0\Lambda^d(E)$ and it follows from [MR, Lemma 4.2.], that

$$C_0\Lambda^d(E) \leq L(\operatorname{cap}(B^n(r),E))^p \tag{4.33}$$

for all $p > 2$. It is immediately seen from the proof of [MR, Lemma 4.2.], that L can be chosen as

$$L = (\max\{CK_1^n 2^n \omega_{n-1}^{-n}, h(r_0)^{1/p}(2K_2)^n \omega_{n-1}^{-n}\})^p,$$

where C and K_2 are positive constants depending only on n, $\omega_{n-1} = H^{n-1}(S(1))$ (see [MR, 3.1]) and

$$K_1 = b_1 \int_0^{r_0} \frac{h(\rho)^{1/(pn)}}{\rho} d\rho = b_1 C_0^{1/(pn)} \int_0^{r_0} \rho^{d/(pn)-1} d\rho = b_1 C_0^{1/(pn)} pnd^{-1} r_0^{d/(pn)},$$

where b_1 is a positive constant depending only on n. Hence $L = D\, C_0\, r_0^d$, where D depends only on n, p, and d. We divide both sides of (4.33) by C_0 and let $\varepsilon \to 0$ to obtain (4.32). □

Unfortunately we do not know whether this result could be modified so as to match the other results of this section. For instance, we do not know whether the right hand side of the inequality (4.32) could be replaced by $Dt^d(\mathrm{cap}(E + B^n(t), E))^p$ for all $t \in (0, d(E))$.

5. The continuum criterion

In this section we study another concept which reflects the thickness of a set at a point in the capacitary sense. First though, we recall the following result which illustrates how the thickness of a set at a point affects the behaviour of the modulus near such a point. In Lemma 5.2 below we prove another result of the same type. A related result for uniformly perfect sets can be found in [As, Theorem 3].

5.1. **Theorem.**[Vu2, 3.5.] *Let $\delta > 0$ and let E be a closed set in $\mathbb{R}^n$ with $\mathrm{cap}(0, E, s) \geq \delta$ for every $s \in (0, 1]$. Then there is a number $d^* > 0$ depending only on δ and n such that if $r \in (0, 1)$ and F_r is a continuum joining $S^{n-1}(r)$ and S^{n-1}, then*

$$\mathsf{M}(\Gamma_r^*) \geq d^* \log \frac{1}{r},$$

where $\Gamma_r^* = \Delta(E, F_r; \mathbb{R}^n)$. □

5.2. **Lemma.** *Let $E \subset \mathbb{R}^n$ be closed with $0 \in E$. Let $r > 0$ and let K be a continuum which joins S^{n-1} and $S^{n-1}(r)$. Assume that there exist constants $b \in (-1, 0)$ and $d > 1$ which satisfy for all $p = 0, 1, 2, \ldots$*

$$\delta_p \geq \max\{d\,\omega_{n-1}(\log(2/s_p))^{1-n}, (\log\log\sqrt{1/r})^b\},$$

where $\delta_p = \mathrm{cap}\,(0, E \setminus B^n(s_p^{3/2}), s_p)$ and $s_p = e^{-e^p}$ for $p = 0, 1, 2, \ldots$. Assume that $(\delta_p)_{p=0}^{\infty}$ is a decreasing sequence. Then

$$\mathsf{M}(\Gamma_r) \geq a_n(1 - 1/d)(\log\log\sqrt{1/r})^{b+1},$$

where $\Gamma_r = \Delta(E, K)$,

$$a_n = 3^{-n} \min\left\{1, \frac{\kappa \log(2 - 1/\sqrt{e})}{\log 2} \min\{1, \kappa\}\right\},$$

$\kappa = c_n(\log 2)^n/\omega_{n-1}$, and c_n is the number in Lemma 2.7.

Proof. Since $p \geq 0$, we always have that $p > \log((\log 2)/(e-2))$, which implies that $2e^{-e^{p+1}} < e^{-2e^p}$. Hence $2s_{p+1} < s_p^2$ for all $p = 0, 1, \ldots$, and the annuli $R(2s_p, s_p^2) = B^n(2s_p) \setminus \overline{B^n(s_p^2)}$ are disjoint. We require that these annuli reach $B^n(r)$, in other words choose p_o sufficiently large to satisfy $r \geq s_{p_0}^2 = e^{-2e^{p_0}}$. The choice $p_0 = \lceil \log \log \sqrt{1/r} \rceil$ fulfills this requirement.

We denote

$$E_j = E \cap R(s_j, s_j^{3/2}),\ K_j = K \cap R(s_j, s_j^{3/2}),$$

$$\Gamma_j = \Delta(E_j, S^{n-1}(2s_j); \overline{B^n(2s_j)}),$$

$$\Gamma_j' = \Delta(E_j, S^{n-1}(2s_j); \overline{B^n(2s_j)} \setminus B^n(s_j^2)),$$

$$\Gamma_j'' = \{\gamma \in \Gamma_j \mid |\gamma| \cap B^n(s_j^2) \neq \emptyset\},$$

and

$$\Delta_j = \Delta(S^{n-1}(s_j^2), S^{n-1}(2s_j))$$

for all j, $0 \leq j \leq p_0$. Then $\mathsf{M}(\Delta(E_j, S^{n-1}(2s_j))) = \mathsf{M}(\Gamma_j)$ by [Vu4, (5.10)]. Since $\Gamma_j'' > \Delta_j$, we have

$$\begin{aligned}\mathsf{M}(\Gamma_j'') &\leq \mathsf{M}(\Delta_j)\\ &= \omega_{n-1}(\log(2/e^{-e^j}))^{1-n}\\ &\leq (1/d)\,\mathsf{M}(\Gamma_j),\end{aligned}$$

where the last inequality follows from the assumptions. Clearly $\Gamma_j = \Gamma_j' \cup \Gamma_j''$ and hence we get

$$\mathsf{M}(\Gamma_j) \leq \mathsf{M}(\Gamma_j') + \mathsf{M}(\Gamma_j'')$$

which leads to

$$\mathsf{M}(\Gamma_j') \geq \mathsf{M}(\Gamma_j) - \mathsf{M}(\Gamma_j'') \geq (1 - 1/d)\,\mathsf{M}(\Gamma_j).$$

By denoting for j, $0 \leq j \leq p_0$,

$$\Gamma^j = \Delta(E_j, K_j; R(2s_j, s_j^2)),$$

we get a family of separate curve families Γ^j.

Denote $F_1 = E_j$, $F_2 = K_j$, $F_3 = F_4 = S^{n-1}(2s_j)$, and $\mathsf{M}(\Gamma_{ij}) = \Delta(F_i, F_j; \overline{B^n(2s_j)} \setminus B^n(s_j^2))$ for $i, j \in \{1, 2, 3, 4\}$. Then $\mathsf{M}(\Gamma_{13})$ and $\mathsf{M}(\Gamma_{24})$ have the upper bound $\omega_{n-1}(\log 2)^{1-n}$. Let $z \in S^{n-1}(2s_j)$ be such a point that $d(K_j, S^{n-1}(2s_j)) = d(K_j, z)$. Denote

$$R = \overline{B^n(z, 2s_j - s_j^{3/2})} \setminus B^n(z, 2s_j - s_j).$$

Using [Vu4, (5.10)] and Lemma 2.7, we get

$$\begin{aligned}\mathsf{M}(\Gamma_{24}) &\geq \mathsf{M}(\Delta(K_j \cap R, S^{n-1}(2s_j) \cap R; R))\\ &\geq c_n \log \frac{2s_j - s_j^{3/2}}{2s_j - s_j}\\ &= c_n \log(2 - 1/\sqrt{e}) > 0\end{aligned}$$

as $2 - 1/\sqrt{e} \approx 1.39 > 1$. From the comparison principle for the modulus (see [Vu4, 5.35.]) and Lemma 2.7 it follows that

$$\begin{aligned}
\mathsf{M}(\Gamma^j) &= \mathsf{M}(\Gamma_{12}) \\
&\geq 3^{-n} \min\{\mathsf{M}(\Gamma_{13}), \mathsf{M}(\Gamma_{24}), c_n \log 2\} \\
&\geq 3^{-n} \min\left\{\mathsf{M}(\Gamma_{13}), \mathsf{M}(\Gamma_{24}) \min\left\{1, \frac{c_n \log 2}{\omega_{n-1}(\log 2)^{1-n}}\right\}\right\} \\
&\geq 3^{-n} \min\left\{\mathsf{M}(\Gamma_{13}), \frac{\mathsf{M}(\Gamma_{13})}{\omega_{n-1}(\log 2)^{1-n}} \mathsf{M}(\Gamma_{24}) \min\{1, \kappa\}\right\} \\
&\geq 3^{-n} \min\left\{1, \frac{\kappa \log(2 - 1/\sqrt{e})}{\log 2} \min\{1, \kappa\}\right\} \mathsf{M}(\Gamma_{13}) \\
&= a_n \, \mathsf{M}(\Gamma'_j) \\
&\geq a_n \, (1 - 1/d) \, \mathsf{M}(\Gamma_j).
\end{aligned}$$

In addition we have that $\Gamma_r \supset \cup_{j=0}^{p_0} \Gamma^j$ and that trivially $\cup_{j=0}^{p_0} \Gamma^j < \Gamma^i$ for all i, $0 \leq i \leq p_0$. From these facts and the assumptions we get

$$\begin{aligned}
\mathsf{M}(\Gamma_r) &\geq \mathsf{M}\left(\bigcup_{j=0}^{p_0} \Gamma^j\right) \\
&\geq \sum_{j=0}^{p_0} \mathsf{M}(\Gamma^j) \\
&\geq a_n \, (1 - 1/d) \sum_{j=0}^{p_0} \mathsf{M}(\Gamma_j) \\
&\geq a_n \, (1 - 1/d)(p_0 + 1) \, \delta_{p_0} \\
&\geq a_n \, (1 - 1/d)(\log\log \sqrt{1/r} + 1)(\log\log \sqrt{1/r})^b \\
&\geq a_n \, (1 - 1/d)(\log\log \sqrt{1/r})^{b+1}. \qquad \square
\end{aligned}$$

Lemma 5.2 can be applied in the study of quasiregular mappings as follows.

5.3. **Corollary.** *Let $K \geq 1$ and let $f : B^n \to \mathbb{R}^n$ be a non-constant K-quasiregular mapping. Let $x \in B^n$ and denote $E = \mathbb{R}^n \setminus fB^n$. Assume that $0 \in E$ and that there exist constants $b \in (-1, 0)$ and $d > 1$ which satisfy for all $p = 0, 1, 2, \ldots$*

$$\delta_p \geq \max\{d\,\omega_{n-1}(\log(2/s_p))^{1-n}, (\log\log \sqrt{1/r})^b\},$$

where $\delta_p = \operatorname{cap}(0, E \setminus B^n(s_p^{3/2}), s_p)$, $s_p = e^{-e^p}$ for $p = 0, 1, 2, \ldots$, and $r = |f(x)|$. Assume that $(\delta_p)_{p=0}^{\infty}$ is a decreasing sequence. Then

$$|f(x)| \geq |f(0)| e^{\alpha(b,d,K,n,x)}, \tag{5.4}$$

where

$$\alpha(b, d, K, n, x) = -2e^{((K/(a_n(1-1/d)))\gamma_n(1/|x|))^{1/(b+1)}}$$

and a_n is the number in Lemma 5.2.

Proof. Let $x \in B^n$, $\varepsilon > 0$. Choose a continuum C joining $f(x)$ and $f(0)$ and satisfying

$$\mu_{fB^n}(f(x), f(0)) \geq \mathsf{M}(\Delta(C, E)) - \varepsilon.$$

Using the fact that f is K-quasiregular, we get by [Vu4, 10.18 (1)] that

$$\mu_{fB^n}(f(x), f(0)) \leq K\mu_{B^n}(x, 0) = K\gamma_n\left(\frac{1}{|x|}\right). \tag{5.5}$$

Assume first that $|f(x)| < |f(0)|$. Then Lemma 5.2 yields

$$\begin{aligned}\mu_{fB^n}(f(x), f(0)) &\geq \mathsf{M}(\Delta(C, E)) - \varepsilon \\ &\geq a_n(1 - 1/d)\left(\log\log\sqrt{\frac{|f(0)|}{|f(x)|}}\right)^{b+1} - \varepsilon.\end{aligned}$$

Letting $\varepsilon \longrightarrow 0$, this together with (5.5) implies

$$\left(\log\log\sqrt{\frac{|f(0)|}{|f(x)|}}\right)^{b+1} \leq \left(\frac{K}{a_n(1 - 1/d)}\right)\gamma_n\left(\frac{1}{|x|}\right)$$

and (5.4) follows. Since

$$e^{\alpha(b,d,K,n,x)} \leq 1,$$

(5.4) holds trivially also if $|f(x)| \geq |f(0)|$. □

The following definition was first given by Martio in [Mar1] for the boundary of a domain in $\overline{\mathbb{R}}^n$. In the case of a closed set in $\mathbb{R}^n$ the converse of this criterion and several related conditions were compared in [MS]. For an example see Theorem 5.8 below.

5.6. **The continuum criterion** $\mathsf{M}(x, C) = \infty$. Let C be a closed set in $\overline{\mathbb{R}}^n$. When K is a continuum in $\mathbb{R}^n$, we denote $\Gamma_K = \Delta(K, C)$. Let $y \in \mathbb{R}^n$ and assume that $x \in C \setminus \{\infty\}$. Then we say that $\mathsf{M}(x, C) = \infty$, if

$$\liminf_{\varepsilon \searrow 0} \mathsf{M}(\Gamma_{C_\varepsilon}) = \infty,$$

where C_ε is any continuum joining y and $S^{n-1}(x, \varepsilon)$. If $x = \infty \in C$, we say that $\mathsf{M}(x, C) = \infty$, if

$$\liminf_{t \nearrow \infty} \mathsf{M}(\Gamma_{C_t}) = \infty,$$

where C_t is a continuum joining y and $S^{n-1}(t)$. It can be shown that this definition is independent of the choice of the point y. If $x \in C$ does not satisfy the condition $\mathsf{M}(x, C) = \infty$, we write $\mathsf{M}(x, C) < \infty$. We note that the condition $\mathsf{M}(x, C) < \infty$ does not mean that $\mathsf{M}(x, C)$ should be understood as a real number depending on x and C.

5.7. **Remarks.** (1) As the notation suggests, the condition $\mathsf{M}(x,C)=\infty$ depends on x. For instance, consider the set

$$C=\left(\bigcup_{k=1}^{\infty}B_k\right)\cup\{0\}\subset\mathbb{R}^2,$$

where $B_k=\overline{B^2(e_1/k,1/(ke^{2\pi k^2}))}$ for $k\in\mathbb{N}\setminus\{0\}$. It follows from example 6.35 that $\mathsf{M}(0,C)<\infty$, but clearly $\mathsf{M}(x,C)=\infty$ for all $x\in C\setminus\{0\}$, since each B_k is a connected set with positive capacity. This also shows that the condition $\mathsf{M}(x,C)=\infty$ is not semicontinuous in the following sense. Let $(x_k)_{k=1}^{\infty}$ be the sequence consisting of the points $x_k=e_1/k\in C$, $k=1,2,\ldots$. Then $x_k\to 0\in C$ as $k\to\infty$ and $\mathsf{M}(x_k,C)=\infty$ for all k but $\mathsf{M}(0,C)<\infty$. On the other hand, the condition $\mathsf{M}(x,C)<\infty$ is not semicontinuous in this sense either. This can be seen by choosing

$$C=\overline{B^2}\cup\bigcup_{k=1}^{\infty}\{x_k\},$$

where $x_k=(1+1/k)e_1$. Now $x_k\to e_1$ as $k\to\infty$ and $\mathsf{M}(x_k,C)<\infty$ for all k but $\mathsf{M}(e_1,C)=\infty$.

(2) It is easy to see that under the following assumption the condition $\mathsf{M}(x,C)=\infty$ either holds or fails for all $x\in C$. Namely, assume that $C\subset\mathbb{R}^n$ is a compact set with $\#C=\infty$ and that there exists a $K>1$ and for every $x,y\in C$ a K-quasiconformal mapping $f:\mathbb{R}^n\to\mathbb{R}^n$ with $fC=C$ and $f(x)=y$. We also observe that conversely, this mapping condition fails to hold for the two choices of the set C in (1).

(3) For a set $E\subset\mathbb{R}^n$ with $\operatorname{cap}E>0$, there need not exist a point $x_0\in E$ with $\mathsf{M}(x_0,E)=\infty$. This fact is illustrated by [Mar2, 4.7.], where Martio constructs an example of a Cantor type compact set $C\subset\mathbb{R}^n$ such that $\operatorname{cap}C>0$ but $\mathsf{M}(x,C)<\infty$ for all $x\in C$. Note that by Theorem 5.8 below, we then also have for each $x\in C$

$$\lim_{t\to 0}\operatorname{cap}(x,C,t)=0.\qquad\square$$

Next we recall three useful results concerning the continuum criterion.

5.8. **Theorem.**[MS, 3.1.(b)] *Let C be closed in $\mathbb{R}^n$ and $x\in C$. If $\mathsf{M}(x,C)<\infty$, then* $\lim_{t\to 0}\operatorname{cap}(x,C,t)=0$. $\square$

5.9. **Theorem.**[Mar1, 3.4.] *Let G be a domain in $\overline{\mathbb{R}}^n$ and suppose that $x\in\partial G$. Then $\mathsf{M}(x,\partial G)<\infty$ if and only if there exists a continuum $C\subset G\cup\{x\}$ such that $x\in C$ and $\mathsf{M}(\Delta(C,\partial G;G))<\infty$.* $\square$

5.10. **Lemma.**[Vu1, 8.3.] *Let G be a domain with $0\in\partial G$ and $\mathsf{M}(0,\partial G)<\infty$. If K is a continuum in $G\cup\{0\}$, $0\in K$, such that $\mathsf{M}(\Delta(K,\complement G;G))<\infty$, then*

$$\lim_{r\to 0}\mathsf{M}(\Gamma_{K_r})=0,$$

where K_r is the 0-component of $K\cap\overline{B^n(r)}$ and $\Gamma_{K_r}=\Delta(K_r,\complement G;G)$. $\square$

5.11. **Theorem.** *If E is a uniformly perfect set, then $\mathsf{M}(x,E)=\infty$ for all $x\in E$.*

Proof. Let E be uniformly perfect. Then there exists, according to Theorem 4.16, a positive constant C such that

$$\operatorname{cap}(x, E, r) \geq C$$

for all $r \in (0, d(E))$ and all $x \in E$.

Assume that there exists such a point $x_0 \in E$ that $\mathsf{M}(x_0, E) < \infty$. Then it follows from Theorem 5.8 that

$$\lim_{r \to 0} \operatorname{cap}(x_0, E, r) = 0.$$

This is a contradiction. □

5.12. **Example.** For $k \in \mathbb{N}$, define $a_k = e^{-e^k}$, $r_k = a_k/10$,

$$E_k = \overline{B^n(a_k e_1, r_k)},$$

and let

$$E = \{0\} \cup \left(\bigcup_{k \in \mathbb{N}} E_k \right).$$

We claim that E is a set with $\mathsf{M}(x, E) = \infty$ for all $x \in E$ but E is not uniformly perfect.

We start by showing that $\mathsf{M}(x, E) = \infty$ for all $x \in E$. Since each of the balls E_k is clearly uniformly perfect, it follows from 5.11 that $\mathsf{M}(x, E_k) = \infty$ for all $k \in \mathbb{N}$ and all $x \in E_k$. This implies that $\mathsf{M}(x, E) = \infty$ for all $x \in E \setminus \{0\}$ and only the point $x = 0$ requires special attention. To show that the condition in 5.6 holds in $x = 0$, we first choose $y \in \mathbb{R}^n$ and $\varepsilon > 0$. Since our interest is in what happens when $\varepsilon \to 0$, we may immediately assume that $\varepsilon < |y|$. Let C_ε be a continuum joining y and $S^{n-1}(\varepsilon)$. Let $k_0 \in \mathbb{N}$ be the smallest such number that $C_\varepsilon \cap S^{n-1}(a_{k_0} + r_{k_0}) \neq \emptyset$ and let $k_1 \in \mathbb{N}$ be the largest such number that $a_k - r_k \geq \varepsilon$ for all $k \leq k_1$. Then $k_0 \leq k_1$. Denote

$$\Gamma_{C_\varepsilon} = \Delta(C_\varepsilon, E)$$

and

$$\Gamma_k = \Delta(E_k, C_\varepsilon; B^n(a_k + r_k) \setminus \overline{B^n(a_k - r_k)})$$

for all k, $k_0 \leq k \leq k_1$. Then the curve families Γ_k are separate and $\Gamma_{C_\varepsilon} < \Gamma_k$ for all k, $k_0 \leq k \leq k_1$. It follows from [Vu4, 5.4.] and Lemma 2.7 that

$$\begin{aligned} \mathsf{M}(\Gamma_{C_\varepsilon}) &\geq \sum_{k=k_0}^{k_1} \mathsf{M}(\Gamma_k) \geq \sum_{k=k_0}^{k_1} c_n \log \frac{a_k + r_k}{a_k - r_k} \\ &= c_n \sum_{k=k_0}^{k_1} \log \frac{11}{9} = (k_1 - k_0) c_n \log \frac{11}{9}. \end{aligned}$$

As $\varepsilon \to 0$, $k_1 \to \infty$ and $\mathsf{M}(\Gamma_{C_\varepsilon}) \to \infty$. Hence $\mathsf{M}(0, E) = \infty$.

It remains for us to demonstrate that E is not uniformly perfect. This fact will follow from Theorem 4.16 if we show that

$$\operatorname{cap}(0, E, a_k - r_k) \to 0 \quad \text{as} \quad k \to \infty.$$

For $k \in \mathbb{N}$, denote

$$\tilde{\Gamma}_k = \Delta\left(\bigcup_{j=k+1}^{\infty} E_j, S^{n-1}(2(a_k - r_k))\right)$$

and

$$\Gamma'_k = \Delta(S^{n-1}(a_{k+1} + r_{k+1}), S^{n-1}(a_k - r_k)).$$

Then $\Gamma'_k < \tilde{\Gamma}_k$ and invoking Lemma 2.6, we find that

$$\begin{aligned} \operatorname{cap}(0, E, a_k - r_k) &= \mathsf{M}(\tilde{\Gamma}_k) \le \mathsf{M}(\Gamma'_k) \\ &= \omega_{n-1}\left(\log\frac{a_k - r_k}{a_{k+1} + r_{k+1}}\right)^{1-n} \\ &= \omega_{n-1}\left(\log\frac{9}{11}e^{e^k(e-1)}\right)^{1-n} \to 0 \end{aligned}$$

as $k \to \infty$. □

5.13. **Comparison of the thickness conditions.** We conclude this section by summarizing the interdependence of the thickness criteria employed in our work. For closed subsets E of $\mathbb{R}^n$, we illustrate the dependencies in the following figure. The diagonal arrows are included for information purposes only, as these already follow from 4.16 with 5.11 or 5.12.

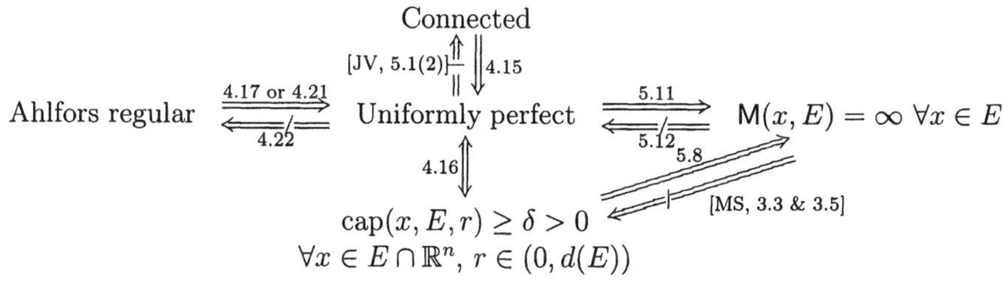

6. Comparison theorems for conformal invariants

One of the key ideas of geometric function theory is conformal invariance. This entails the following. Firstly, one should strive for results where this invariance is apparent. Secondly, the geometric quantities used should have this invariance property. Thirdly, the results should be natural in the sense that in special cases one should be able to obtain the previously known results. Well-known examples of conformal invariants of complex analysis are the harmonic measure, the modulus of a curve family, and the hyperbolic metric of a plane domain.

In this section we study two conformal invariants associated with a pair of points in a domain $G \subsetneq \mathbb{R}^n$. Later on we shall assume that G fulfills some additional assumptions by imposing conditions on the metric properties of ∂G.

We first recall the definitions of some metrics which are constantly used in the study of quasiconformal and quasiregular mappings and which we shall apply in our study of conformal invariants.

6.1. **The metrics j_G and k_G.** For an open set $G \subsetneq \mathbb{R}^n$, define

$$j_G(x,y) = \log\left(1 + \frac{|x-y|}{\min\{d(x),\, d(y)\}}\right), \quad x,y \in G.$$

It was shown in [Se, 2.2.] that j_G is a metric.

Let G be a proper subdomain of $\mathbb{R}^n$. The *quasihyperbolic metric* k_G is defined by

$$k_G(x,y) = \inf_{\gamma \in \Gamma_{xy}} \int_\gamma \frac{|dz|}{d(z)}, \quad x,y \in G,$$

where Γ_{xy} is the family of all rectifiable curves in G joining x and y. The *quasihyperbolic ball* $D_k(x,r)$ is the set $\{z \in G \mid k_G(x,z) < r\}$, when $x \in G$ and $r > 0$. By [Vu4, (3.9)], we have the inclusions

(6.2) $$B^n(x, rd(x)) \subset D_k(x,M) \subset B^n(x, Rd(x)),$$

where $r = 1 - e^{-M}$ and $R = e^M - 1$. If $G = H^n$, the numbers r and R are the best possible. It is well known (see [Vu4, 3.3.]), that

(6.3) $$k_G(x,y) \geq j_G(x,y) \geq \left|\log \frac{d(x)}{d(y)}\right|, \quad x,y \in G.$$

6.4. **The conformal invariants λ_G and μ_G.** Let G be a proper subdomain of $\overline{\mathbb{R}}^n$. Denote $C_z = \gamma_z[0,1)$, where $\gamma_z : [0,1) \to G$ is a curve satisfying $z \in \gamma_z[0,1)$ and $\gamma_z(t) \to \partial G$, when $t \to 1$. Then, for x, $y \in G$, $x \neq y$, we define

$$\lambda_G(x,y) = \inf_{C_x, C_y} \mathsf{M}(\Delta(C_x, C_y; G)).$$

We also define for x, $y \in G$

$$\mu_G(x,y) = \inf_{C_{xy}} \mathsf{M}(\Delta(C_{xy}, \partial G; G)),$$

where the infimum is taken over all continua C_{xy} such that $C_{xy} = \gamma[0,1]$ and γ is a curve with $\gamma(0) = x$ and $\gamma(1) = y$.

It follows from Definition 2.8 that λ_G and μ_G are invariant under conformal mappings of G and quasi-invariant under quasiconformal maps.

In a general domain G, there exist no known expressions neither for λ_G nor for μ_G in terms of well-known simple functions. For $G = B^n$, the formulae

(6.5) $$\mu_{B^n}(x,y) = 2^{n-1}\tau_n\left(\frac{1}{\sinh^2(\frac{1}{2}\rho_{B^n}(x,y))}\right) = \gamma_n\left(\frac{1}{\tanh(\frac{1}{2}\rho_{B^n}(x,y))}\right)$$

and

(6.6) $$\lambda_{B^n}(x,y) = \frac{1}{2}\tau_n\left(\sinh^2\left(\frac{1}{2}\rho_{B^n}(x,y)\right)\right)$$

hold by [Vu4, 8.6.]. Here ρ_{B^n} denotes the hyperbolic metric of B^n (see Section 2 of [Vu4]) for which it is well known that

$$\sinh^2\left(\frac{1}{2}\rho_{B^n}(x,y)\right) = \frac{|x-y|^2}{(1-|x|^2)(1-|y|^2)}.$$

It follows that for $x, y \in B^n$,

$$\lambda_{B^n}(x,y) = \frac{1}{2}\tau_n\left(\frac{|x-y|^2}{(1-|x|^2)(1-|y|^2)}\right). \tag{6.7}$$

When $G = \mathbb{R}^n \setminus \{0\}$, we have by [Vu4, (8.23)]

$$\lambda_{\mathbb{R}^n\setminus\{0\}}(x,y) = \min\{p(r_x(y)), p(r_y(x))\}, \tag{6.8}$$

where r_z, for $z \in \mathbb{R}^n \setminus \{0\}$, is a similarity map with $r_z(0) = 0$ and $r_z(z) = e_1$ and p is the function defined in (2.23). It was proved in [Vu4, 8.25.] that if G is a proper subdomain of $\mathbb{R}^n$ and if $x, y \in G$ with $x \neq y$, then

$$\lambda_G(x,y) \le \inf_{z\in\partial G} \lambda_{\mathbb{R}^n\setminus\{z\}}(x,y) \le 4\tau_n\left(\frac{|x-y|}{\min\{d(x),d(y)\}}\right). \tag{6.9}$$

For $G = \mathbb{R}^n \setminus \{0\}$ a stronger upper bound is given by [Vu5, 1.7.]: for $x, y \in G$, $x \neq y$,

$$\tau_n\left(\frac{|x-y|}{\min\{|x|,|y|\}}\right) \le \lambda_G(x,y) \le \tau_n\left(\frac{|x-y| + \big|\,|x|-|y|\,\big|}{2\min\{|x|,|y|\}}\right). \tag{6.10}$$

For several other estimates for μ_G and λ_G we refer the reader to Section 8 of [Vu4].

It can be shown that μ_G is a metric if $\operatorname{cap}\partial G > 0$, see [Vu4, 8.1.]. In this case μ_G is called the *modulus metric* or *conformal metric* of G. If G is a proper subdomain of $\mathbb{R}^n$, then $\lambda_G^{-1/n}$ is a metric. This was proved in [LF]. In [AVV, 16.1.] it was shown that $\lambda_{B^n}^{-p}$ is a metric if and only if $p \in (0, 1/(n-1)]$. By the Riemann mapping theorem and the conformal invariance of λ_G it is seen immediately that for a simply connected proper subdomain G of $\mathbb{R}^2$, λ_G^{-p} is a metric if and only if $p \in (0,1]$. For a generalization, see [Fe].

When G is a proper subdomain of $\mathbb{R}^n$, $a \in G$, and $t > 0$, we denote

$$D_{\lambda^{-1}}(a,t) = \{b \in G \,|\, \lambda_G(a,b) > t\} \text{ and } D_\mu(a,t) = \{b \in G \,|\, \mu_G(a,b) < t\}.$$

6.11. **Theorem.** *Let G be a proper subdomain of $\mathbb{R}^n$ and let $t > 0$. Denote $c_1 = 1/(1+\tau_n^{-1}(t/4))$, $c_2 = \sqrt{\tau_n^{-1}(2t)/(1+\tau_n^{-1}(2t))}$ and $c_3 = \tau_n^{-1}(t/4)$. Then the inclusions*

$$D_{\lambda^{-1}}(a,t) \subset \{z \in G \,|\, d(z) > c_1 d(a)\}, \tag{6.12}$$

$$D_{\lambda^{-1}}(a,t) \supset B^n(a, c_2 d(a)) \supset D_k(a, \log(c_2+1)), \tag{6.13}$$

and

$$D_{\lambda^{-1}}(a,t) \subset B^n(a, c_3 d(a)) \cap G \tag{6.14}$$

are valid for all $a \in G$. If in addition $t > 4\tau_n(1)$, then we have that

$$B^n(a, c_3 d(a)) \subset D_k(a, \log(1/(1-c_3))). \tag{6.15}$$

Proof. Let $a, z \in G$, $a \neq z$ with $\lambda_G(a,z) > t$. We may assume that $d(a) > d(z)$. The triangle inequality yields

$$|z-a| \geq d(a) - d(z) > 0.$$

It follows from (6.9) that

$$\lambda_G(a,z) \leq 4\tau_n(|a-z|/\min\{d(a),d(z)\}) = 4\tau_n(|a-z|/d(z)).$$

Since τ_n is a strictly decreasing homeomorphism, this leads to

$$\frac{d(a)-d(z)}{d(z)} < \tau_n^{-1}(t/4)$$

which is equivalent to

$$d(z) > \frac{d(a)}{1+\tau_n^{-1}(t/4)}.$$

We proceed to the proof of the left side inclusion in (6.13). Let a, $z \in G$, $|z-a| < c_2 d(a)$. Then, since $c_2 < 1$, we have $z \in B^n(a, d(a))$ and [Vu4, 8.8. (1)] implies

$$\lambda_G(a,z) \geq \frac{1}{2}\tau_n\left(\frac{r^2}{1-r^2}\right),$$

where $r = |z-a|/d(a)$. Since the function τ_n is strictly decreasing and the function $s \mapsto s^2/(1-s^2)$ is strictly increasing for $s > 0$, we get

$$\lambda_G(a,z) > \frac{1}{2}\tau_n\left(\frac{c_2^2}{1-c_2^2}\right) = t.$$

The right side inclusion follows from (6.2).

To prove the inclusion (6.14), we again apply (6.9) to obtain

$$\lambda_G(a,z) \leq 4\tau_n\left(\frac{|z-a|}{d(a)}\right).$$

This means that, with the assumption $t < \lambda_G(a,z)$, we have

$$|z-a| < \tau_n^{-1}(t/4)d(a).$$

Since $D_{\lambda^{-1}}(a,t) \subset G$, the inclusion (6.14) holds.

Here we have used (6.9) repeatedly but it is worth noting that if $n = 2$, we could instead compare λ_G with $\lambda_{\mathbb{R}^2\setminus\{0\}}$ and use [Vu4, (8.23)] with [BV, 1.1.] to obtain a more accurate upper bound for λ_G in terms of τ_2. We omit the details.

Finally, (6.15) follows directly from (6.2) after we notice that the condition $t > 4\tau_n(1)$ implies that $c_3 < 1$ and hence that the ball $B^n(a, c_3 d(a))$ is included in G. □

Although λ_G^{-1} is usually not a metric (see 6.4), the following topological result holds.

6.16. **Lemma.** *Let G be a proper subdomain of $\mathbb{R}^n$. The balls $D_{\lambda^{-1}}(a,t)$, $a \in G$, $t > 0$, form a basis of a topology of G.*

Proof. Obviously the balls $D_{\lambda^{-1}}(a,t)$, $a \in G$, $t > 0$, form a covering of G. We have to show that if $a_1, a_2 \in G$, $t_1, t_2 > 0$ and if $x \in \cap_{i=1}^2 D_{\lambda^{-1}}(a_i,t_i)$, then there exist $a_3 \in G$, $t_3 > 0$ satisfying

$$x \in D_{\lambda^{-1}}(a_3,t_3) \subset \bigcap_{i=1}^{2} D_{\lambda^{-1}}(a_i,t_i).$$

For this, it suffices, according to Theorem 6.11, to find an $r > 0$ satisfying

$$B^n(x,r) \subset \bigcap_{i=1}^{2} D_{\lambda^{-1}}(a_i,t_i).$$

Such an r always exists if we show that the balls $D_{\lambda^{-1}}(a,t)$, $a \in G$, $t > 0$, are open in the relative Euclidean topology. This in turn follows from the fact that for a fixed $z \in G$, the function $\lambda_G(z,\cdot)$ is continuous, which we now prove.

Fix $x, y_1 \in G$. We show that the function $y \mapsto \lambda_G(x,y)$ is continuous at y_1. Let $y_2 \in G$. We may assume that $y_2 \in B^n(y_1, d(y_1))$ so that $J_{y_1y_2} \subset G$. Let C, D be continua joining x, y_1 to ∂G, respectively. Then

$$\begin{aligned}\lambda_G(x,y_2) &\le \mathsf{M}(\Delta(C, J_{y_1y_2} \cup D; G)) \\ &\le \mathsf{M}(\Delta(C, J_{y_1y_2}; G)) + \mathsf{M}(\Delta(C, D; G)).\end{aligned}$$

Taking the infimum over all such continua C and D, we get

$$|\lambda_G(x,y_2) - \lambda_G(x,y_1)| \le \inf_C \mathsf{M}(\Delta(C, J_{y_1y_2}; G)) \to 0, \quad \text{as } y_2 \to y_1.$$

This completes the proof. □

6.17. **Corollary.** *The balls $D_{\lambda^{-1}}(a,t)$, $a \in G$, $t > 0$, form a basis of the relative Euclidean topology of G.*

Proof. Let $x \in D_{\lambda^{-1}}(a,t)$, $a \in G$, $t > 0$. We must find such a Euclidean ball $B \subset G$ that $x \in B \subset D_{\lambda^{-1}}(a,t)$. But as was seen in the proof of Lemma 6.16, $D_{\lambda^{-1}}(a,t) \subset G$ is open in the relative Euclidean topology of G. Hence we may choose $B = B^n(x,r)$ with a sufficiently small r.

Conversely, let $x \in B^n(z,r) \subset G$, $r > 0$. Since $B^n(z,r)$ is open, we find such an $r' > 0$ that $B^n(x,r') \subset B^n(z,r)$. Then $r' < d(x)$, let $r' = cd(x)$, where $0 < c < 1$. By Theorem 6.11 we have $B^n(x,r') \supset D_{\lambda^{-1}}(x,t)$ with $t = 4\tau_n(c)$. Hence

$$x \in D_{\lambda^{-1}}(x, 4\tau_n(c)) \subset B^n(x,r') \subset B^n(z,r)$$

as required. □

Next we turn to the metric μ_G. The following theorems imply that when G has a thick boundary, the topology defined by μ_G is equivalent with the relative Euclidean topology and the topology defined by the quasihyperbolic metric of G.

6.18. **Theorem.** *Let G be a proper subdomain of $\mathbb{R}^n$ and assume that G has a connected, nondegenerate boundary. Let $t > 0$ and denote $d_1 = \tau_n^{-1}(t)/(1+\tau_n^{-1}(t))$, $d_2 = 1/\gamma_n^{-1}(t)$ and $d_3 = 1/\tau_n^{-1}(t)$. Then, for all $a \in G$, the following inclusions hold*

$$D_\mu(a,t) \subset \{z \in G \,|\, d(z) > d_1 d(a)\}, \tag{6.19}$$

$$D_\mu(a,t) \supset B^n(a, d_2 d(a)) \supset D_k(a, \log(d_2+1)), \tag{6.20}$$

$$D_\mu(a,t) \subset B^n(a, d_3 d(a)) \cap G. \tag{6.21}$$

If in addition $t < \tau_n(1)$, then

$$B^n(a, d_3 d(a)) \subset D_k(a, \log(1/(1-d_3))). \tag{6.22}$$

The numbers d_1, d_2 and d_3 are the best possible for these inclusions.

Proof. Let $a, z \in G$ satisfy $\mu_G(a,z) < t$. Let C_{az} be a continuum joining a and z in G. Since $d_1 < 1$, we may assume that $d(z) < d(a)$ in proving (6.19). We use spherical symmetrization with center at z and the fact that ∂G is connected, to obtain

$$\mathsf{M}(\Delta(C_{az}, \partial G; G)) \geq \tau_n\left(\frac{d(z)}{|z-a|}\right).$$

Taking the infimum over all such continua C_{az}, we get

$$\mu_G(a,z) \geq \tau_n\left(\frac{d(z)}{|z-a|}\right).$$

By the triangle inequality, we have

$$d(a) \leq |z-a| + d(z),$$

which leads to

$$t > \mu_G(a,z) \geq \tau_n\left(\frac{d(z)}{|z-a|}\right) \geq \tau_n\left(\frac{d(z)}{d(a)-d(z)}\right).$$

Since τ_n is a strictly decreasing homeomorphism, we get

$$\frac{d(z)}{d(a)-d(z)} > \tau_n^{-1}(t),$$

or equivalently,

$$d(z) > \left(\frac{\tau_n^{-1}(t)}{1+\tau_n^{-1}(t)}\right) d(a).$$

By choosing $G = \mathbb{C} \setminus (-\infty, -1]$, $a = 1$, and $z = 0$, we see that d_1 is the best possible constant for (6.19).

Next we assume that a, $z \in G$ and that $|z-a| < d_2 d(a)$. Then, since $\gamma_n^{-1}(t) > 1$, we have $z \in B^n(a, d(a)) \subset G$. We denote

$$\Gamma_J = \Delta(J_{az}, \partial G; G),$$

$$\Gamma = \Delta(J_{az}, S^{n-1}(a, d(a)); \overline{B^n(a, d(a))}),$$

and

$$\tilde{\Gamma} = \Delta([z', \infty), S^{n-1}; \mathbb{R}^n \setminus B^n),$$

where $z' = (d(a)/|z-a|)e_1$. Since J_{az} is a continuum which joins a and z and since $\Gamma < \Gamma_J$, we have that

$$\mu_G(a, z) \le \mathsf{M}(\Gamma_J) \le \mathsf{M}(\Gamma). \tag{6.23}$$

Using Möbius transformations, we get

$$\mathsf{M}(\Gamma) = \mathsf{M}(\tilde{\Gamma}) = \gamma_n\left(\frac{d(a)}{|z-a|}\right), \tag{6.24}$$

and since $|z-a| < d_2 d(a)$ and γ_n is a strictly decreasing homeomorphism, it follows that

$$\gamma_n\left(\frac{d(a)}{|z-a|}\right) < \gamma_n\left(\frac{1}{d_2}\right) = t. \tag{6.25}$$

Combining the inequalities (6.23), (6.24), and (6.25), we get

$$\mu_G(a, z) < t,$$

which proves the left side of (6.20). By choosing $G = B^n$, $a = 0$, and $z = d_2 e_1$, we can show that d_2 is the best possible constant in this inclusion. The right side inclusion of (6.20) follows from (6.2), which is sharp in the case $G = H^n$.

For the inclusion (6.21), let again $a, z \in G$ satisfy $\mu_G(a, z) < t$ and let C_{az} be a continuum joining a and z in G. Since ∂G is connected, spherical symmetrization with center at a gives

$$\mathsf{M}(\Delta(C_{az}, \partial G; G)) \ge \tau_n\left(\frac{d(a)}{|z-a|}\right).$$

Taking the infimum over all such continua C_{az} leads to

$$t > \mu_G(a, z) \ge \tau_n\left(\frac{d(a)}{|z-a|}\right),$$

which is equivalent to

$$|z-a| < \frac{d(a)}{\tau_n^{-1}(t)}.$$

Now (6.21) follows from this and the fact that $D_\mu(a, t) \subset G$. We choose $G = \mathbb{C} \setminus (-\infty, -1]$, $a = 0$, and $z = 1$ to see that d_3 is the best constant for this inclusion.

Note that if $t < \tau_n(1)$, then $B^n(a, d_3 d(a)) \subset G$. Then the inclusion (6.22) follows immediately from (6.2), which also gives the best possible constant for this inclusion in the case $G = H^n$. □

Next we relax the assumption on the boundary of G. Connectedness is not required, but we do need to assume that a weaker thickness condition holds. Note that (6.20) holds regardless of the thickness of the boundary of G.

6.26. **Theorem.** *Let G be a proper subdomain of $\mathbb{R}^n$ and assume that there exists such a strictly decreasing function $h : (0,1) \to (0,\infty)$ that for every point $x \in \partial G$ and for every continuum $K \subset G$ satisfying $K \cap S^{n-1}(x,\alpha) \neq \emptyset$ and $K \cap S^{n-1}(x,\beta) \neq \emptyset$ for $0 < \alpha < \beta < d(G)$, the inequality*

$$\mathsf{M}(\Delta(K, \partial G; \mathbb{R}^n)) \geq h\left(\frac{\alpha}{\beta}\right)$$

holds. Let $t > 0$ and denote $d_1' = h^{-1}(t)$, $d_2' = 1/\gamma_n^{-1}(t)$ and $d_3' = 1 + 1/h^{-1}(t)$. Then, for all $a \in G$, the following inclusions are valid

(6.27) $$D_\mu(a,t) \subset \{z \in G \,|\, d(z) > d_1' d(a)\},$$

(6.28) $$D_\mu(a,t) \supset B^n(a, d_2' d(a)) \supset D_k(a, \log(d_2' + 1)),$$

(6.29) $$D_\mu(a,t) \subset B^n(a, d_3' d(a)) \cap G.$$

Proof. Let $a, z \in G$. We choose such points a_0 and z_0 on ∂G that $|a - a_0| = d(a)$ and $|z - z_0| = d(z)$. Also, we let C_{az} be a continuum which joins a and z in G and denote $\Gamma_{az} = \Delta(C_{az}, \partial G; G)$. By [Vu4, (5.10)] we have that $\mathsf{M}(\Gamma_{az}) = \mathsf{M}(\Delta(C_{az}, \partial G; \mathbb{R}^n))$.

For the proof of (6.27), assume that $\mu_G(a,z) < t$. As $d_1' < 1$, we may assume that $d(z) < d(a)$. It is clear that $|a - z_0| \geq d(a)$ and it follows that $|a - z_0| > d(z)$. Since C_{az} joins $S^{n-1}(z_0, d(z))$ and $S^{n-1}(z_0, |a - z_0|)$, we get

$$\mathsf{M}(\Gamma_{az}) \geq h\left(\frac{d(z)}{|a - z_0|}\right) \geq h\left(\frac{d(z)}{d(a)}\right).$$

We take the infimum over all such continua C_{az} to get

$$t > \mu_G(a,z) \geq h\left(\frac{d(z)}{d(a)}\right),$$

which leads to

$$d(z) > h^{-1}(t)\, d(a).$$

The inclusions in (6.28) are just a restatement of (6.20).

It remains to prove (6.29). Let again $\mu_G(a,z) < t$. If $|z - a| \leq d(a)$, we are done, since $d_3' > 1$. Hence we assume that $|z - a| > d(a)$. There are two cases to be considered.

Case 1: $|z - a_0| > d(a)$.
Now $z \notin \overline{B^n(a_0, d(a))}$ and using the triangle inequality, we get

$$|z - a_0| \geq |z - a| - d(a) > 0.$$

It follows that

$$\mathsf{M}(\Gamma_{az}) \geq h\left(\frac{d(a)}{|z - a_0|}\right) \geq h\left(\frac{d(a)}{|z - a| - d(a)}\right).$$

Case 2: $|z - a_0| \leq d(a)$.
In this case $d(z) \leq d(a)$, which implies that $|z_0 - a| \geq d(z)$. If $|z_0 - a| = d(z)$, then we have that $d(a) \leq |z_0 - a| = d(z) \leq d(a)$, which means that $d(z) = d(a)$. Then $|z - a_0| \leq d(a) = d(z)$, which yields

$|z-a_0| = d(z)$. Hence $z, a \in S^{n-1}(a_0, d(a))$, which means that $|z-a| \le 2d(a) < d_3' d(a)$, as $d_3' > 2$. Therefore we may assume that $|z_0 - a| > d(z)$. Then the triangle inequality gives

$$|z_0 - a| \ge |z-a| - d(z) \ge |z-a| - d(a) > 0$$

and we get

$$\mathsf{M}(\Gamma_{az}) \ge h\left(\frac{d(z)}{|z_0-a|}\right) \ge h\left(\frac{d(a)}{|z-a|-d(a)}\right).$$

In both cases 1 and 2 we take the infimum over all such continua C_{az} to get

$$t > \mu_G(a,z) \ge h\left(\frac{d(a)}{|z-a|-d(a)}\right).$$

Consequently, we have that

$$|z-a| < \left(1 + \frac{1}{h^{-1}(t)}\right) d(a).$$

We conclude the proof by noting that $D_\mu(a,t) \subset G$. □

6.30. **Corollary.** *Let G be a proper subdomain of $\mathbb{R}^n$ and assume that ∂G is uniformly perfect. Let $t > 0$ and denote $\tilde{d_1} = 1/e^{t/d^*}$, $\tilde{d_2} = 1/\gamma_n^{-1}(t)$ and $\tilde{d_3} = 1 + e^{t/d^*}$, where d^* is the constant in Theorem 5.1. Then, for all $a \in G$, the following inclusions are valid*

(6.31) $$D_\mu(a,t) \subset \{z \in G \,|\, d(z) > \tilde{d_1} d(a)\},$$

(6.32) $$D_\mu(a,t) \supset B^n(a, \tilde{d_2} d(a)) \supset D_k(a, \log(\tilde{d_2}+1)),$$

(6.33) $$D_\mu(a,t) \subset B^n(a, \tilde{d_3} d(a)) \cap G.$$

Proof. Let $x \in \partial G$ and let $0 < \alpha < \beta < d(G)$. Assume that K is a continuum in G with $K \cap S^{n-1}(x,\alpha) \ne \emptyset \ne K \cap S^{n-1}(x,\beta)$. Let G' and K' be the images of G and K, respectively, under the Möbius mapping $z \mapsto z/\beta$. Since the modulus of a ring is a conformal invariant, it is clear by definition 4.15 that $\partial G'$ is also uniformly perfect. Hence we get by Theorems 4.16 and 5.1, that

$$\mathsf{M}(\Delta(K, \partial G; \mathbb{R}^n)) = \mathsf{M}(\Delta(K', \partial G'; \mathbb{R}^n)) \ge d^* \log \frac{1}{\alpha/\beta}$$

for $0 < \alpha < \beta < d(G)$. This means that we can apply Theorem 6.26 with $h(r) = d^* \log(1/r)$ and the inclusions in (6.31), (6.32), and (6.33) follow. □

We have shown that uniform perfectness of the boundary of G implies the compatibility of μ_G and k_G and the fact that $\mathsf{M}(x, \partial G) = \infty$ for all $x \in \partial G$. The next theorem shows that the continuum criterion on ∂G is necessary for the compatibility of the said metrics to hold.

6.34. **Theorem.** *Let G be a proper subdomain of $\mathbb{R}^n$ and assume that $\mathsf{M}(x_0, \partial G) < \infty$ for some point $x_0 \in \partial G$. Then, for every $R > 0$, there exists such a point $a \in G$ that*

$$D_\mu(a,R) \not\subset D_k(a,r) \quad \text{for all } r > 0.$$

Proof. Since $\mathsf{M}(x_0, \partial G) < \infty$, there exists, by Theorem 5.9, a nondegenerate continuum $K \subset G \cup \{x_0\}$ satisfying $x_0 \in K$ and $\mathsf{M}(\Gamma_K) < \infty$, where $\Gamma_K = \Delta(K, \partial G; G)$. It follows from Lemma 5.10 that

$$\lim_{r \to 0} \mathsf{M}(\Gamma_{K_r}) = 0,$$

where K_r is the x_0-component of $K \cap \overline{B^n(x_0, r)}$.

Let $R > 0$. We choose an $r_0 > 0$ which satisfies $\mathsf{M}(\Gamma_{K_r}) < R$ for all $r \leq r_0$ and denote by C the x_0-component of $K \cap B^n(x_0, r_0)$. Let a be a point in $C \cap G$. We show next that $x_0 \in \overline{D_\mu(a, R)}$. Let $\varepsilon > 0$. We may assume that $\varepsilon < d(C)$. Choose a point $z \in (C \cap B^n(x_0, \varepsilon)) \setminus \{x_0\}$ which belongs to the a-component C' of $C \setminus B^n(x_0, \varepsilon')$, where $\varepsilon' = |z - x_0| > 0$. Then C' is connected, bounded and closed, since $\complement C' = \complement C \cup B^n(x_0, \varepsilon')$ is open. Hence C' is a continuum. Moreover $\Gamma_{C'} \subset \Gamma_C$, which leads to

$$\mu_G(a, z) \leq \mathsf{M}(\Gamma_{C'}) \leq \mathsf{M}(\Gamma_C) \leq \mathsf{M}(\Gamma_{K_{r_0}}) < R.$$

Consequently $z \in D_\mu(a, R)$, which proves that $x_0 \in \overline{D_\mu(a, R)} \cap \partial G$.

On the other hand, (6.3) implies that

$$k_G(a, y) \geq \left| \log \frac{d(a)}{d(y)} \right| \to \infty, \quad \text{as } y \to \partial G,\ y \in G.$$

Hence $\overline{D_k(a, r)} \cap \partial G = \emptyset$ for all $r > 0$, which means that for all $r > 0$

$$D_\mu(a, R) \not\subset D_k(a, r). \qquad \square$$

6.35. **Example.** For every positive integer k, define

$$B_k = \overline{B^2(e_1/k, 1/(ke^{2\pi k^2}))} \text{ and } S_k = S^1(e_1/k, 1/(ke^{2\pi k^2})).$$

It is easy to see that the sets B_k are disjoint. Define

$$G = \mathbb{R}^2 \setminus \left(\left(\bigcup_{k=1}^{\infty} B_k \right) \cup \{0\} \right).$$

Then G is a domain in $\mathbb{R}^2$ and

$$\partial G = \left(\bigcup_{k=1}^{\infty} S_k \right) \cup \{0\}.$$

Let $C = [-e_1, 0] \subset G \cup \{0\}$ and denote

$$\Gamma_C = \Delta(C, \partial G; G),$$

$$\Gamma_k = \Delta(C, S_k; G)$$

and

$$\Gamma_k' = \Delta(S^1(e_1/k, 1/k), S_k; G)$$

for $k = 1, 2, \ldots$. Now $\Gamma'_k < \Gamma_k$ and using Lemma 2.6, we get

$$\begin{aligned} \mathsf{M}(\Gamma_k) &\le \mathsf{M}(\Gamma'_k) \\ &\le \mathsf{M}(\Delta(S^1(e_1/k, 1/k), S_k; \mathbb{R}^2)) \\ &= \omega_1 \left(\log \frac{1/k}{1/(ke^{2\pi k^2})} \right)^{1-2} \\ &= 1/k^2. \end{aligned}$$

Since $\Gamma_C = \bigcup_{k=1}^{\infty} \Gamma_k$, we get

$$\mathsf{M}(\Gamma_C) \le \sum_{k=1}^{\infty} \mathsf{M}(\Gamma_k) \le \sum_{k=1}^{\infty} \frac{1}{k^2} < \infty.$$

Now it follows from Theorem 5.9, that $\mathsf{M}(0, \partial G) < \infty$. For $r > 0$, we denote $a_r = -re_1$, $K_r = [-re_1, 0]$ and $\Gamma_{K_r} = \Delta(K_r, \partial G; G)$. Then we have, according to Lemma 5.10, that

$$\lim_{r \to 0} \mathsf{M}(\Gamma_{K_r}) = 0.$$

Let $\varepsilon > 0$ and let $r_0 > 0$ be so small that $\mathsf{M}(\Gamma_{K_r}) < \varepsilon$ for $r \le r_0$. We will show that $0 \in \overline{D_\mu(a_{r_0}, \varepsilon)}$. Let $\delta > 0$. Choose a point $y \in K_{r_0} \cap (B^2(\delta) \setminus \{0\})$. Then

$$\begin{aligned} \mu_G(a_{r_0}, y) &\le \mathsf{M}(\Delta([a_{r_0}, y], \partial G; G)) \\ &\le \mathsf{M}(\Gamma_{K_{r_0}}) \\ &< \varepsilon, \end{aligned}$$

which means that $y \in D_\mu(a_{r_0}, \varepsilon)$. Hence $B^2(\delta) \cap D_\mu(a_{r_0}, \varepsilon) \ne \emptyset$, which means that $0 \in \overline{D_\mu(a_{r_0}, \varepsilon)}$.

Conclusion: We have constructed such a domain G in $\mathbb{R}^2$ that $\operatorname{cap}(\partial G) > 0$ and for all $\varepsilon > 0$ there exists a point $a \in G$ satisfying

$$\overline{D_\mu(a, \varepsilon)} \cap \partial G \ne \emptyset. \qquad \square$$

7. Removing a point

It is evident from the definition of λ_G that adding new points, even isolated ones, to the boundary of G will affect the value of $\lambda_G(x, y)$ for fixed points $x, y \in G$, neither of which is included in the new boundary points. We study this phenomenon in some simple cases where a point is removed from the domain G. In the first case we assume that G is a QED domain.

7.1. **QED domains.** Let $c \in (0, 1]$. A closed set E in $\overline{\mathbb{R}}^n$ is called a *c-quasiextremal distance set* or a *c-QED set*, if

$$\mathsf{M}(\Delta(F_1, F_2; \complement E)) \ge c\,\mathsf{M}(\Delta(F_1, F_2))$$

for all disjoint continua $F_1, F_2 \subset \complement E$. A domain $G \subset \overline{\mathbb{R}}^n$ is a c-QED *domain*, if $\complement G$ is a c-QED set.

7.2. **Theorem.** *Let G be a proper subdomain of $\mathbb{R}^n$ and assume that G is c-QED for some $c \in (0,1]$. Let $x_1 \in G$ and denote $G_1 = G \setminus \{x_1\}$. Then for each $\alpha \in (0,1)$ there exists a constant d depending on n, α and c and satisfying*

$$\frac{\lambda_{G_1}(x,y)}{\lambda_G(x,y)} \geq d$$

for all distinct $x, y \in G \setminus B^n(x_1, \alpha\, d(x_1))$.

Proof. We denote $d(z) = d(z, \partial G)$, when $z \in G$ and $d_1(z) = d(z, \partial G_1)$, when $z \in G_1$. We may assume that $d_1(x) \leq d_1(y)$. From (6.9) we get $\lambda_G(x,y) \leq 4\tau_n(m)$, where $m = |x-y|/d(x)$ and from [Vu4, 8.29.] we get $\lambda_{G_1}(x,y) \geq 2^{1-n} c\, \tau_n(m_1)$, where $m_1 = |x-y|/d_1(x)$. Note that $d_1(x) \leq d(x)$. Since $x \in G \setminus B^n(x_1, \alpha\, d(x_1))$, we notice that

$$d(x) = d(x, \partial G) \leq (1 + 1/\alpha) d(x, \partial G_1) = (1 + 1/\alpha) d_1(x),$$

which leads to

$$m_1 \leq (1 + 1/\alpha) m.$$

The function τ_n is strictly decreasing, so

$$\frac{\lambda_{G_1}(x,y)}{\lambda_G(x,y)} \geq \left(\frac{c}{2^{n+1}}\right)\left(\frac{\tau_n(m_1)}{\tau_n(m)}\right) \geq \left(\frac{c}{2^{n+1}}\right)\left(\frac{\tau_n((1+1/\alpha)m)}{\tau_n(m)}\right).$$

Using (2.17), we get $\tau_n((1+1/\alpha)m) > (1/\sqrt{1+1/\alpha})\, \tau_n(m)$, which leads to the conclusion

$$\frac{\lambda_{G_1}(x,y)}{\lambda_G(x,y)} \geq \frac{c}{2^{n+1}\sqrt{1+1/\alpha}}. \qquad \square$$

It follows from the definition of λ_G that if G_1 and G_2 are proper subdomains of $\mathbb{R}^n$ with $G_1 \subset G_2$, then $\lambda_{G_1}(x,y) \leq \lambda_{G_2}(x,y)$ for all $x, y \in G_1$. One may wish to ask when equality or strict inequality holds. The next theorem gives a sufficient condition for equality in the case $G_1 = B^n \setminus \{0\}$ and $G_2 = B^n$.

7.3. **Theorem.** *Let $G = B^n \setminus \{0\}$ and let $x, y \in G$ with $|x-y| \geq \delta > 0$. Then, if $\min\{|x|,|y|\} \in (r_1, 1)$ with $r_1 = (\sqrt{\delta^4 + 64} - \delta^2)/8$, we have that*

$$\lambda_G(x,y) = \lambda_{B^n}(x,y).$$

Proof. Assume that the claim does not hold. Then $\lambda_G(x,y) < \lambda_{B^n}(x,y)$ and by relabeling if necessary, we may assume that there exist sets C_x and C_y as in definition 6.4 with C_x joining x and 0 and C_y joining y and S^{n-1} in G with

$$\mathsf{M}(\Delta(C_x, C_y; G)) < \lambda_{B^n}(x,y).$$

By (6.7) we have that

$$\lambda_{B^n}(x,y) = \frac{1}{2}\tau_n\left(\frac{|x-y|^2}{(1-|x|^2)(1-|y|^2)}\right) \leq \frac{1}{2}\tau_n\left(\frac{\delta^2}{(1-|x|^2)(1-|y|^2)}\right).$$

Let f be a Möbius transformation $B^n \to H^n$ and let $\tilde{f}$ be the extension of f to $\overline{\mathbb{R}}^n$ by reflection in S^{n-1}, see [Vä2, p. 119]. By [Vu4, 5.22] we get

$$\begin{aligned} \mathsf{M}(\Delta(C_x, C_y; G)) &= \mathsf{M}(\Delta(C_x, C_y; B^n)) \\ &= \mathsf{M}(\Delta(f(C_x), f(C_y); H^n)) \\ &\geq \frac{1}{2}\mathsf{M}(\Delta(f(C_x), f(C_y))) \\ &= \frac{1}{2}\mathsf{M}(\Delta(\tilde{f}^{-1}f(C_x), \tilde{f}^{-1}f(C_y))) \\ &= \frac{1}{2}\mathsf{M}(\Delta(C_x, C_y)). \end{aligned}$$

We may add to C_x the point 0 and to C_y a point in S^{n-1} (see 6.4), making these sets continua, without affecting the modulus of $\Delta(C_x, C_y)$. For simplicity, we still denote these larger sets by C_x and C_y. Let C_x^* and C_y^* be the spherical symmetrizations of C_x and C_y, respectively, with respect to 0 and the x_1-axis. Furthermore, denote by $\tilde{C}_x^*$ and $\tilde{C}_y^*$ the images under the inversion in S^{n-1} of C_x^* and C_y^*, respectively. Then, by Theorem 2.11 and by the conformal invariance of the modulus, we have that

$$\begin{aligned} \mathsf{M}(\Delta(C_x, C_y)) &\geq \mathsf{M}(\Delta(C_x^*, C_y^*)) \\ &= \mathsf{M}(\Delta(\tilde{C}_x^*, \tilde{C}_y^*)) \\ &= \tau_n\left(\frac{1+1/|x|}{1/|y|-1}\right) \\ &= \tau_n\left(\frac{(|x|+1)|y|}{(1-|y|)|x|}\right). \end{aligned}$$

Now it suffices to show that under the assumptions of the theorem, we have the inequality

$$\frac{(|x|+1)|y|}{(1-|y|)|x|} < \frac{\delta^2}{(1-|x|^2)(1-|y|^2)},$$

or equivalently,

$$\frac{(1+|x|)(1+|y|)|y|}{|x|} < \frac{\delta^2}{1-|x|^2}.$$

Assuming that $0 < r < |x|, |y| < 1$, the left hand side is $< 4/r$. Hence it is enough to show that

$$\frac{4}{r} \leq \frac{\delta^2}{1-r^2},$$

which is equivalent to

$$4r^2 + \delta^2 r - 4 \geq 0.$$

But using the quadratic formula, this is seen to hold for $r \geq r_1$ with equality if $r = r_1$. This is a contradiction. We also note that since $\delta^4 + 8^2 < (\delta^2 + 8)^2$, we have $r_1 < 1$ as is required for the claim to be sensible. This concludes the proof. □

7.4. **Remark.** The condition $|x-y| \geq \delta$ is needed in Theorem 7.3. This can be seen as follows. Let $G = B^2 \setminus \{0\}$ and let $x, y \in G$ with $x = te_1$ and $y = se_1$, where $0 < t < s < 1$.

We claim that $\lambda_G(x,y) < \lambda_{B^2}(x,y)$. To prove this, we first notice that by Lemma 3.17 we have

$$\lambda_G(x,y) \leq \tau_2\left(\frac{(s-t)(1-st)}{t(1-s)^2}\right)$$

and by (6.7),

$$\lambda_{B^2}(x,y) = \frac{1}{2}\tau_n\left(\frac{(s-t)^2}{(1-t^2)(1-s^2)}\right).$$

Denote

$$\alpha(s,t) = \frac{(s-t)^2}{(1-t^2)(1-s^2)}.$$

We use (3.7) to obtain

$$\frac{1}{2}\tau_2(\alpha(s,t)) = \tau_2((\sqrt{\alpha(s,t)} + \sqrt{\alpha(s,t)+1})^4 - 1).$$

A lengthy but elementary calculation leads to

$$(\sqrt{\alpha(s,t)} + \sqrt{\alpha(s,t)+1})^4 - 1 = \frac{4(s-t)(1-st)}{(1+t)^2(1-s)^2}.$$

Hence it suffices to show that

$$\frac{(s-t)(1-st)}{t(1-s)^2} > \frac{4(s-t)(1-st)}{(1+t)^2(1-s)^2},$$

which is, after canceling terms, equivalent to $4t < (1+t)^2$ and further to $(t-1)^2 > 0$. But this last inequality holds since $t \in (0,1)$. This concludes the proof of the claim.

Now if $\delta > 0$ and $r_1 = (\sqrt{\delta^4 + 64} - \delta^2)/8$, we may take $r_1 < t < s < 1$. Then $|x|, |y| \in (r_1, 1)$ but by Claim we have $\lambda_G(x,y) < \lambda_{B^2}(x,y)$. It is easy to see that $|x-y| < 1 - r_1 < \delta$. □

It is natural to also ask when the converse case occurs, that is, to find conditions on $|x|$ and $|y|$ which would imply that $\lambda_G(x,y) < \lambda_{B^n}(x,y)$. At first glance, this problem seems to be more difficult, the reason being that there are few suitable upper bounds for $\lambda_G(x,y)$. We begin our investigation in the plane, where we have more tools at our disposal. One of these is the function λ defined in (3.8).

7.5. **Theorem.** *Let $x, y \in G = B^2 \setminus \{0\}$, $|x| \leq |y|$, $x \neq y$ and let $\delta = |x-y| > 0$ be fixed so that $\delta < \frac{2\lambda(2)}{\sqrt{1+4\lambda(2)^2}}$. If*

$$|y| < \sqrt{1 - \frac{4\delta^2\lambda(2)^2}{4\lambda(2)^2 - \delta^2}} \quad \text{and} \quad |x| < \sqrt{\frac{1-|y|^2}{4\lambda(2)^2 + 1 - |y|^2}},$$

then

$$\lambda_G(x,y) < \lambda_{B^2}(x,y).$$

Proof. By (6.7) we have that

$$\lambda_{B^2}(x,y) = \frac{1}{2}\tau_2\left(\frac{\delta^2}{(1-|x|^2)(1-|y|^2)}\right).$$

Using (6.10), we get

$$\lambda_G(x,y) \le \lambda_{\mathbb{R}^2\setminus\{0\}}(x,y) \le \tau_2\left(\frac{|x-y|}{2\min\{|x|,|y|\}}\right) = \tau_2\left(\frac{\delta}{2|x|}\right).$$

Furthermore, by (3.9) we have

$$\frac{1}{2}\tau_2(t) \ge \tau_2(\lambda(2)\max\{t^2, t^{1/2}\}), \quad t \in (0,\infty),$$

so we set

$$\frac{\delta}{2\lambda(2)|x|} = \max\{t^2, t^{1/2}\}.$$

There are two cases to be considered.

Case 1: $\frac{\delta}{2\lambda(2)|x|} \ge 1$.

In this case $|x| \le \frac{\delta}{2\lambda(2)}$ and $t^2 = \frac{\delta}{2\lambda(2)|x|}$, giving $t = \sqrt{\frac{\delta}{2\lambda(2)|x|}}$. Hence the claim holds if

$$\frac{1}{2}\tau_2\left(\sqrt{\frac{\delta}{2\lambda(2)|x|}}\right) < \frac{1}{2}\tau_2\left(\frac{\delta^2}{(1-|x|^2)(1-|y|^2)}\right).$$

Since τ_2 is a decreasing homeomorphism, this is equivalent to

$$\sqrt{\frac{\delta}{2\lambda(2)|x|}} > \frac{\delta^2}{(1-|x|^2)(1-|y|^2)},$$

which after squaring transforms to the equivalent form

$$\frac{|x|}{(1-|x|^2)^2} < \frac{(1-|y|^2)^2}{2\delta^3\lambda(2)}.$$

Here

$$\frac{|x|}{(1-|x|^2)^2} \le \frac{\delta/(2\lambda(2))}{(1-\delta^2/(4\lambda(2)^2))^2} = \frac{8\delta\lambda(2)^3}{(4\lambda(2)^2-\delta^2)^2}.$$

Hence the claim holds if

$$\frac{(1-|y|^2)^2}{2\delta^3\lambda(2)} > \frac{8\delta\lambda(2)^3}{(4\lambda(2)^2-\delta^2)^2}$$

which is to say

$$|y| < \sqrt{1 - \frac{4\delta^2\lambda(2)^2}{4\lambda(2)^2-\delta^2}}.$$

This upper bound makes sense only if

$$4\lambda(2)^2 - \delta^2 - 4\delta^2\lambda(2)^2 > 0,$$

or equivalently,

$$\delta < \frac{2\lambda(2)}{\sqrt{1+4\lambda(2)^2}}.$$

Case 2: $\frac{\delta}{2\lambda(2)|x|} < 1$.
Now $|x| > \frac{\delta}{2\lambda(2)}$ and $t^{1/2} = \frac{\delta}{2\lambda(2)|x|}$ which implies $t = \frac{\delta^2}{4\lambda(2)^2|x|^2}$. In this case the claim is valid if

$$\frac{1}{2}\tau_2\left(\frac{\delta^2}{4\lambda(2)^2|x|^2}\right) < \frac{1}{2}\tau_2\left(\frac{\delta^2}{(1-|x|^2)(1-|y|^2)}\right)$$

which is equivalent to

$$\frac{\delta^2}{4\lambda(2)^2|x|^2} > \frac{\delta^2}{(1-|x|^2)(1-|y|^2)}$$

and furthermore to

$$|x| < \sqrt{\frac{1-|y|^2}{4\lambda(2)^2+1-|y|^2}}.$$

Taking into account the lower bound $|x| > \frac{\delta}{2\lambda(2)}$, this only makes sense if

$$\frac{\delta}{2\lambda(2)} < \sqrt{\frac{1-|y|^2}{4\lambda(2)^2+1-|y|^2}},$$

which is equivalent to

$$|y| < \sqrt{1 - \frac{4\delta^2\lambda(2)^2}{4\lambda(2)^2-\delta^2}}.$$

The proof is complete. □

7.6. **Remark.** We recall from (3.10) that

$$\lambda(2) = 16 + 12\sqrt{2} \approx 32.97,$$

so that

$$4\lambda(2)^2 = 2176 + 1536\sqrt{2} \approx 4348.23,$$

and

$$\frac{2\lambda(2)}{\sqrt{1+4\lambda(2)^2}} \approx 0.999885. \qquad \square$$

The proof of Theorem 7.5 was based on the estimate (6.10), which is valid in $\mathbb{R}^n$, $n \geq 2$, but we were forced to restrict ourselves to the plane as the function λ, defined in the plane, was used for the estimation of the Teichmüller capacity function. In the plane we can also approach this question by using Lemma 3.17 as in the following theorem.

7.7. **Theorem.** *If $y \in G = B^2 \setminus \{0\}$ with $|y| = t \in (0,1)$, then*

$$\lambda_G(x,y) < \lambda_{B^2}(x,y)$$

whenever $x \in G$ with $|x| < \varepsilon$, where

$$\varepsilon = \frac{1}{2}\left(\frac{\sigma+1+(1-\sigma)t^2}{t} - \sqrt{\left(\frac{t^2(\sigma-1)-\sigma-1}{t}\right)^2 - 4}\right)$$

and

$$\sigma = \tau_2^{-1}\left(\frac{1}{2}\tau_2\left(\frac{4t^2}{(1-t^2)^2}\right)\right).$$

Proof. It follows from Lemma 3.17 that for $x \in G$, $|x| \leq t$,

$$\lambda_G(x,y) \leq \tau_2\left(\frac{(t-|x|)(1-t|x|)}{|x|(1-t^2)}\right).$$

On the other hand, by (6.7),

$$\lambda_{B^2}(x,y) = \frac{1}{2}\tau_2\left(\frac{|x-y|^2}{(1-|x|^2)(1-|y|^2)}\right) \geq \frac{1}{2}\tau_2\left(\frac{(2t)^2}{(1-t^2)^2}\right).$$

In order to have the claim hold, it is enough to have that

$$\tau_2\left(\frac{(t-|x|)(1-t|x|)}{|x|(1-t^2)}\right) < \frac{1}{2}\tau_2\left(\frac{4t^2}{(1-t^2)^2}\right),$$

which is equivalent to

$$\frac{(t-|x|)(1-t|x|)}{|x|(1-t^2)} > \sigma$$

and further to

(7.8)
$$|x|^2 + \frac{t^2(\sigma-1)-\sigma-1}{t}|x| + 1 > 0.$$

Since $t \in (0,1)$ and $\sigma > 0$, we have that

$$t^2(\sigma-1)-\sigma-1 = \sigma(t^2-1)-t^2-1 < 0.$$

Using this information and the quadratic formula, we see rather easily that the left hand side of (7.8) has two distinct positive roots, namely

$$|x| = \frac{1}{2}\left(\frac{\sigma+1+(1-\sigma)t^2}{t} \pm \sqrt{\left(\frac{t^2(\sigma-1)-\sigma-1}{t}\right)^2 - 4}\right).$$

Since their product is equal to the constant term 1, the smaller root, namely $|x| = \varepsilon$, is in $(0,1)$. The claim follows. $\square$

In order to answer the question at hand also in $\mathbb{R}^n$, $n \geq 2$, we first prove an upper bound for $\lambda_{B^n\setminus\{0\}}$. We note that this upper bound is quite close in form, but improves, the upper bound which follows from (6.10).

7.9. **Theorem.** *Let $G = B^n \setminus \{0\}$, $x, y \in G$, $x \neq y$. Then*

$$\begin{aligned}\lambda_G(x,y) &\leq \tau_n\left(\frac{|x-y| + \big||y|-|x|\big|}{2\min\{|x|(1-|y|), |y|(1-|x|)\}}\right)\\ &< \sqrt{2}\,\tau_n\left(\frac{|x-y| + \big||y|-|x|\big|}{\min\{|x|(1-|y|), |y|(1-|x|)\}}\right).\end{aligned}$$

Proof. By (2.28), we have that

$$p(z) \leq \tau_n\left(\frac{|z| + |z-e_1| - 1}{2}\right)$$

for $z \in \mathbb{R}^n \setminus [0, e_1]$. By Möbius invariance this implies that if $a, b, c, d \in \mathbb{R}^n$ are distinct points, then there exist continua E and F with $a, b \in E$ and $c, d \in F$ such that

$$\mathsf{M}(\Delta(E,F)) \leq \tau_n(\xi),$$

where

$$\xi = \frac{|a,b,c,d| + |b,a,c,d| - 1}{2}.$$

Let $x, y \in G$ be distinct points. Choosing above $a = 0$, $b = x$, $c = y$, and $d \in S^{n-1}$ with $|y-d| = 1-|y|$, we get

$$1 + 2\xi = \frac{|y||x-d|}{|x||y-d|} + \frac{|x-y||d|}{|x||y-d|} = \frac{|y||x-d| + |x-y|}{|x|(1-|y|)} \geq \frac{|y|(1-|x|) + |x-y|}{|x|(1-|y|)},$$

which implies that

$$\xi \geq \frac{|x-y| + |y| - |x|}{2|x|(1-|y|)}.$$

Interchanging the roles of x and y above, we get

$$\xi \geq \frac{|x-y| + |x| - |y|}{2|y|(1-|x|)}.$$

Hence we have that

$$\xi \geq \frac{|x-y| + \big|\,|y|-|x|\,\big|}{2\min\{|x|(1-|y|), |y|(1-|x|)\}}$$

and it follows that

$$\begin{aligned}\lambda_G(x,y) &\leq \tau_n(\xi) \leq \tau_n\left(\frac{|x-y| + \big|\,|y|-|x|\,\big|}{2\min\{|x|(1-|y|), |y|(1-|x|)\}}\right) \\ &< \sqrt{2}\,\tau_n\left(\frac{|x-y| + \big|\,|y|-|x|\,\big|}{\min\{|x|(1-|y|), |y|(1-|x|)\}}\right).\end{aligned}$$

Here the last inequality follows from (2.16). □

7.10. **Theorem.** *Let $x, y \in G = B^n \setminus \{0\}$, $x \neq y$, and let $|y| = t \in (0,1)$. Then, if*

$$|x| < \frac{t}{\sigma(1-t)+1},$$

where

$$\sigma = \tau_n^{-1}\left(\frac{1}{2}\tau_n\left(\frac{4t^2}{(1-t^2)^2}\right)\right),$$

we have that

$$\lambda_G(x,y) < \lambda_{B^n}(x,y).$$

Proof. Using Theorem 7.9, the fact that $|x| \leq |y|$, and the triangle inequality, we get

$$\begin{aligned}\lambda_G(x,y) &\leq \tau_n\left(\frac{|x-y| + \big|\,|y|-|x|\,\big|}{2\min\{|x|(1-|y|), |y|(1-|x|)\}}\right) \\ &= \tau_n\left(\frac{|x-y| + |y|-|x|}{2|x|(1-|y|)}\right) \\ &\leq \tau_n\left(\frac{2(|y|-|x|)}{2|x|(1-|y|)}\right) \\ &= \tau_n\left(\frac{t-|x|}{|x|(1-t)}\right).\end{aligned}$$

On the other hand, by (6.7) we have that

$$\lambda_{B^n}(x,y) = \frac{1}{2}\tau_n\left(\frac{|x-y|^2}{(1-|x|^2)(1-|y|^2)}\right) \geq \frac{1}{2}\tau_n\left(\frac{(2t)^2}{(1-t^2)^2}\right).$$

Hence the claim holds if

$$\tau_n\left(\frac{t-|x|}{|x|(1-t)}\right) < \frac{1}{2}\tau_n\left(\frac{4t^2}{(1-t^2)^2}\right)$$

which in turn holds if and only if

$$t - |x| > \sigma|x|(1-t),$$

or equivalently,

$$|x| < \frac{t}{\sigma(1-t)+1}. \qquad \square$$

Another natural problem is to find conditions on $x, y \in G = B^n \backslash \{0\}$ under which $\lambda_G(x,y) > c\lambda_{B^n}(x,y)$ for some $c \in (0,1)$. The following theorem provides a solution to this problem.

7.11. **Theorem.** *Let $c \in (0, 2^{1-n})$ and denote $a = c^{1/(1-n)}$. If $x, y \in G = B^n \setminus \{0\}$ with*

$$|x-y| > \min\{\alpha(x,y)^{1/(2(1-1/a))}\delta(x,y)^{1/(1-a)}, 4^{n-1}c^2\alpha(x,y)\delta(x,y)^{-1}\}, \tag{7.12}$$

where $\alpha(x,y) = (1-|x|^2)(1-|y|^2)$ and $\delta(x,y) = \min\{|x|, 1-|x|, |y|, 1-|y|\}$, then

$$\lambda_G(x,y) > c\,\lambda_{B^n}(x,y).$$

Proof. We recall that by (6.7),

$$\lambda_{B^n}(x,y) = \frac{1}{2}\tau_n\left(\frac{|x-y|^2}{(1-|x|^2)(1-|y|^2)}\right) = \frac{1}{2}\tau_n\left(\frac{|x-y|^2}{\alpha(x,y)}\right).$$

Let $x, y \in G$ and let C_x, C_y be continua joining x and y to ∂G, respectively. Then we have that

$$\min\{d(C_x), d(C_y)\} \geq \delta(x,y).$$

Since B^n and H^n are Möbius equivalent, we may use [Vu4, 5.22.] to get

$$\mathsf{M}(\Delta(C_x, C_y; G)) = \mathsf{M}(\Delta(C_x, C_y; B^n)) \geq \frac{1}{2}\mathsf{M}(\Delta(C_x, C_y)).$$

Next we apply Lemma 2.32(2) to obtain

$$\mathsf{M}(\Delta(C_x, C_y)) \geq 2^{1-n}\tau_n\left(\frac{d(C_x, C_y)}{\min\{d(C_x), d(C_y)\}}\right) \geq 2^{1-n}\tau_n\left(\frac{|x-y|}{\delta(x,y)}\right).$$

Since these inequalities hold for all such continua C_x and C_y, it suffices to show that

$$\tau_n\left(\frac{|x-y|}{\delta(x,y)}\right) > 2^{n-1}c\,\tau_n\left(\frac{|x-y|^2}{\alpha(x,y)}\right). \tag{7.13}$$

Since $c < 2^{1-n}$, we have $a/2 > 1$ and (2.20) implies

$$\tau_n\left(\frac{|x-y|}{\delta(x,y)}\right) \geq \left(\frac{a}{2}\right)^{1-n}\tau_n\left(\left(\frac{|x-y|}{\delta(x,y)}\right)^{2/a}\right) = 2^{n-1}c\,\tau_n\left(\left(\frac{|x-y|}{\delta(x,y)}\right)^{2/a}\right).$$

Hence the claim holds if

$$\left(\frac{|x-y|}{\delta(x,y)}\right)^{2/a} < \frac{|x-y|^2}{\alpha(x,y)},$$

which is equivalent to

$$|x-y| > \alpha(x,y)^{1/(2(1-1/a))}\delta(x,y)^{1/(1-a)}.$$

This gives the first term in the minimum in (7.12).

On the other hand, we may continue after (7.13) by applying (2.16) to get

$$\tau_n\left(\frac{|x-y|}{\delta(x,y)}\right) > \sqrt{(2^{n-1}c)^2}\,\tau_n\left((2^{n-1}c)^2\frac{|x-y|}{\delta(x,y)}\right) = 2^{n-1}c\,\tau_n\left(4^{n-1}c^2\frac{|x-y|}{\delta(x,y)}\right).$$

Note that since $c < 2^{1-n}$, we have $4^{n-1}c^2 < 1$ as required. The claim holds if

$$4^{n-1}c^2\frac{|x-y|}{\delta(x,y)} < \frac{|x-y|^2}{\alpha(x,y)},$$

which is equivalent to

$$|x-y| > \frac{4^{n-1}c^2\alpha(x,y)}{\delta(x,y)},$$

where the right hand side term is the second term in the minimum in (7.12). □

7.14. **Remark.** In the proof of Theorem 7.11 we can apply [Vu4, 7.38] instead of [Vu4, 7.39]. A reasoning similar to the one above then shows that the condition (7.12) may be replaced by the condition

$$\frac{|x-y|}{\delta(x,y)^2} + \frac{1}{\delta(x,y)} < \max\left\{\frac{1}{4c^2}\frac{|x-y|}{\alpha(x,y)}, \frac{|x-y|^{2c-1}}{\alpha(x,y)^c}\right\}.$$

We omit the details of this variation of the proof. □

For the domains $G = B^n$ and $G = \mathbb{R}^n \setminus \{0\}$ one can find expressions or inequalities for $\lambda_G(x,y)$ in terms of the quantity $|x-y|/\min\{d(x),d(y)\}$. One may ask whether similar results hold also for other domains. We shall answer this question in the plane for the domain $G = B^2 \setminus \{0\}$. We start by recalling two estimates for λ_G.

7.15. **Theorem.**[LeVu, 1.6.] *For $G = B^2 \setminus \{0\}$ and $x, y \in G$, $x \neq y$, the following inequality holds:*

(7.16)
$$\lambda_G(x,y) \le \tau_2\left(\frac{|x-y|}{4\min\{|x|(1-|y|),|y|(1-|x|)\}}\right) < C^2\tau_2\left(\frac{|x-y|}{\min\{|x|(1-|y|),|y|(1-|x|)\}}\right),$$

where $C < 1.172$. □

It follows from (7.16) that for $x, y \in G = B^2 \setminus \{0\}$, $x \neq y$, the inequality

(7.17)
$$\lambda_G(x,y) \le C^2\tau_2\left(\frac{|x-y|}{\min\{|x|,1-|y|\}}\right)$$

holds.

7.18. **Theorem.**[ALV, 1.3.] *For distinct $x, y \in G = B^2 \setminus \{0\}$, $|x| \le |y|$, we have*

$$\lambda_G(x,y) \le \tau_2\left(\frac{|x-y|(1+|x|)^2}{2|x|(1-\frac{1}{2}|x-y|)^2}\right) \le \tau_2\left(\frac{|x-y|(1+|x|)^2}{2|x|}\right). \qquad \square \tag{7.19}$$

The inequalities (7.19) and (2.16) imply that for $x, y \in G = B^2 \setminus \{0\}$, $x \ne y$, $|x| \le |y|$,

$$\lambda_G(x,y) \le \tau_2\left(\frac{|x-y|}{2|x|}\right) \le \sqrt{2}\,\tau_2\left(\frac{|x-y|}{|x|}\right).$$

Hence we get for $x, y \in G$, $x \ne y$, that

$$\lambda_G(x,y) \le \sqrt{2}\,\tau_2\left(\frac{|x-y|}{\min\{|x|,|y|\}}\right)$$

and for $x, y \in G$, $x \ne y$, $|x|, |y| \le 1/2$, that

$$\lambda_G(x,y) \le \sqrt{2}\,\tau_2\left(\frac{|x-y|}{\min\{|x|, 1-|x|, |y|, 1-|y|\}}\right). \tag{7.20}$$

7.21. **Theorem.** *Let $x, y \in G = B^2 \setminus \{0\}$, $x \ne y$ with $|x-y| \ge \delta$ for some $\delta \in (0,2)$. Then*

$$\lambda_G(x,y) \le C\tau_2\left(\frac{|x-y|}{\min\{|x|, 1-|x|, |y|, 1-|y|\}}\right)$$

with

$$C = \max\left\{\sqrt{2}, \frac{2(1-r_1)}{\sqrt{2r_1-1}}, \frac{1}{\sqrt{\delta}}\right\},$$

where $r_1 = (\sqrt{\delta^4+64} - \delta^2)/8$.

Proof. By relabeling if necessary, we may assume that $|x| \le |y|$. Since $|x-y| \ge \delta$, we have by Theorem 7.3 that $\lambda_G(x,y) = \lambda_{B^2}(x,y)$ whenever $|x|, |y| > r_1$. Elementary calculus shows that the function $f(z) = (\sqrt{z^4+64} - z^2)/8$, $z \in (0,2)$, is strictly decreasing. As $\delta \in (0,2)$ it follows that $r_1 > (\sqrt{2^4+64} - 2^2)/8 = (\sqrt{5}-1)/2 > 1/2$. Hence we may denote

$$\begin{aligned} A &= B^2 \setminus \overline{B^2(r_1)}, \\ B &= \overline{B^2(r_1)} \setminus \overline{B^2(1/2)}, \\ C &= \overline{B^2(1/2)}. \end{aligned}$$

The following cases may occur.

Case 1: $x, y \in A$.
In this case, by Theorem 7.3 and (6.7),

$$\lambda_G(x,y) = \lambda_{B^2}(x,y) = \frac{1}{2}\tau_2\left(\frac{|x-y|^2}{(1-|x|^2)(1-|y|^2)}\right).$$

Since $1-|x|, 1-|y|<1$ and $1+|x|, 1+|y|<2$, we get

$$\begin{aligned}(1-|x|^2)(1-|y|^2) &= (1-|x|)(1+|x|)(1-|y|)(1+|y|)\\ &< 4(1-|x|)(1-|y|)\\ &\le 4\min\{1-|x|, 1-|y|\}.\end{aligned}$$

Since $\delta/4 \in (0,1/2)$ and $|x-y|\ge\delta$, we may apply (2.16) to get

$$\begin{aligned}\frac{1}{2}\tau_2\left(\frac{|x-y|^2}{(1-|x|^2)(1-|y|^2)}\right) &\le \frac{1}{2}\tau_2\left(\frac{\delta}{4}\frac{|x-y|}{\min\{1-|x|,1-|y|\}}\right)\\ &< \frac{1}{2}\cdot\frac{1}{\sqrt{\delta/4}}\tau_2\left(\frac{|x-y|}{\min\{1-|x|,1-|y|\}}\right)\\ &= \frac{1}{\sqrt{\delta}}\tau_2\left(\frac{|x-y|}{\min\{|x|,1-|x|,|y|,1-|y|\}}\right).\end{aligned}$$

Here we also used the facts that $1-|x|\le|x|$ and $1-|y|\le|y|$.

Case 2: $x\in B, y\in A\cup B$.
Now $\min\{|x|,1-|y|\}=1-|y|=\min\{|x|,1-|x|,|y|,1-|y|\}$, so that (7.17) yields

$$\lambda_G(x,y)\le C_1^2\tau_2\left(\frac{|x-y|}{\min\{|x|,1-|y|\}}\right)=C_1^2\tau_2\left(\frac{|x-y|}{\min\{|x|,1-|x|,|y|,1-|y|\}}\right).$$

Here and in the remainder of the proof C_1 denotes the constant C in Theorem 7.15.

Case 3: $x\in C, y\in A$.
If $|x|>1-|y|$, we get as in Case 2, that

$$\lambda_G(x,y)\le C_1^2\tau_2\left(\frac{|x-y|}{\min\{|x|,1-|x|,|y|,1-|y|\}}\right).$$

Assume that $|x|\le 1-|y|$. Then $|x-y|\le|x|+|y|\le 1$ and $|y|-|x|\ge r_1-1/2$, implying $|y|-|x|\ge(r_1-1/2)|x-y|$. As $|x|\le 1/2$ and $|y|<1$, we have $1-|x||y|\ge 1-1/2=1/2$. Furthermore, by the fact that $1-|y|\le 1-r_1$ and the assumption, we get $|x|(1-|y|)^2\le|x|(1-r_1)^2=(1-r_1)^2\min\{|x|,1-|x|,|y|,1-|y|\}$. Now we use these facts together with Lemma 3.17 to obtain

$$\begin{aligned}\lambda_G(x,y) &\le \mathsf{M}(\Delta([0,x],[y,y/|y|];B^2))\\ &\le \mathsf{M}(\Delta([0,x],[|y|x/|x|,x/|x|];B^2))\\ &= \tau_2\left(\frac{(|y|-|x|)(1-|x||y|)}{|x|(1-|y|)^2}\right)\\ &\le \tau_2\left(\frac{(1/2)(r_1-1/2)|x-y|}{(1-r_1)^2\min\{|x|,1-|x|,|y|,1-|y|\}}\right)\\ &= \tau_2\left(\frac{r_1-1/2}{2(1-r_1)^2}\cdot\frac{|x-y|}{\min\{|x|,1-|x|,|y|,1-|y|\}}\right).\end{aligned}$$

Here $(r_1-1/2)/(2(1-r_1)^2) < 1$ if and only if $4r_1^2-10r_1+5>0$ if and only if $r_1 \notin [(5-\sqrt{5})/4,(5+\sqrt{5})/4] = [0.6909\ldots,1.8090\ldots]$. Hence if $r_1 \in ((\sqrt{5}-1)/2,(5-\sqrt{5})/4)$, we have by (2.16) that

$$\tau_2\left(\frac{r_1-1/2}{2(1-r_1)^2}\cdot\frac{|x-y|}{\min\{|x|,1-|x|,|y|,1-|y|\}}\right) < \frac{2(1-r_1)}{\sqrt{2r_1-1}}\tau_2\left(\frac{|x-y|}{\min\{|x|,1-|x|,|y|,1-|y|\}}\right).$$

On the other hand, if $r_1 \in [(5-\sqrt{5})/4,1)$, then $(r_1-1/2)/(2(1-r_1)^2) \geq 1$ and therefore

$$\tau_2\left(\frac{r_1-1/2}{2(1-r_1)^2}\cdot\frac{|x-y|}{\min\{|x|,1-|x|,|y|,1-|y|\}}\right) \leq \tau_2\left(\frac{|x-y|}{\min\{|x|,1-|x|,|y|,1-|y|\}}\right).$$

We conclude that in this case,

$$\lambda_G(x,y) \leq \max\left\{1,\frac{2(1-r_1)}{\sqrt{2r_1-1}}\right\}\tau_2\left(\frac{|x-y|}{\min\{|x|,1-|x|,|y|,1-|y|\}}\right).$$

Case 4: $x\in C, y\in B$.
If $|x| \leq 1-|y|$, then $\min\{|x|,1-|y|\} = |x| = \min\{|x|,1-|x|,|y|,1-|y|\}$. If, on the other hand, $|x| > 1-|y|$, then $\min\{|x|,1-|y|\} = 1-|y| = \min\{|x|,1-|x|,|y|,1-|y|\}$. Using (7.17), we obtain

$$\lambda_G(x,y) \leq C_1^2\tau_2\left(\frac{|x-y|}{\min\{|x|,1-|y|\}}\right) = C_1^2\tau_2\left(\frac{|x-y|}{\min\{|x|,1-|x|,|y|,1-|y|\}}\right).$$

Case 5: $x,y\in C$.
Now we have by (7.20) that

$$\lambda_G(x,y) \leq \sqrt{2}\,\tau_2\left(\frac{|x-y|}{\min\{|x|,1-|x|,|y|,1-|y|\}}\right).$$

By Cases $1-5$ we have

$$\lambda_G(x,y) \leq \max\left\{\frac{1}{\sqrt{\delta}},C_1^2,1,\frac{2(1-r_1)}{\sqrt{2r_1-1}},\sqrt{2}\right\}\tau_2\left(\frac{|x-y|}{\min\{|x|,1-|x|,|y|,1-|y|\}}\right).$$

According to Theorem 7.15, $C_1 < 1.172$, which implies that $C_1^2 < 1.374$. Hence $C_1^2 < \sqrt{2} \approx 1.414$. The claim follows. □

7.22. **Remark.** In the above theorem, $\delta \in (0,2)$ implies

$$r_1 \in \left(\frac{\sqrt{2^4+64}-2^2}{8},1\right) = \left(\frac{\sqrt{5}-1}{2},1\right)$$

and hence that

$$\frac{2(1-r_1)}{\sqrt{2r_1-1}} \in \left(0,\frac{3-\sqrt{5}}{\sqrt{\sqrt{5}-2}}\right) = (0,1.5723\ldots).$$

On the other hand,

$$\frac{1}{\sqrt{\delta}} \in \left(\frac{1}{\sqrt{2}},\infty\right) = (0.7071\ldots,\infty).$$ □

References

[AB] Ahlfors, L. and A. Beurling: Conformal invariants and function-theoretic null-sets. - Acta Math. 83 (1950), 101–129.

[ALV] Anderson, G. D., M. Lehtinen, and M. Vuorinen: Conformal invariants in the punctured unit disk. - Ann. Acad. Sci. Fenn. Ser. A I Math. 19 (1994), 133–146.

[AVV] Anderson, G. D., M. K. Vamanamurthy, and M. Vuorinen: Conformal Invariants, Inequalities, and Quasiconformal Maps. - J. Wiley, 1997.

[As] Aseev, V. V.: Continuity of conformal capacity for condensers with uniformly perfect plates. - Siberian Math. J. 40 (1999), 205–213.

[Bea1] Beardon, A. F.: On the Hausdorff dimension of general Cantor sets. - Proc. Camb. Phil. Soc. 61 (1965), 679–694.

[Bea2] Beardon, A. F.: The Geometry of Discrete Groups. - Graduate Texts in Math., Vol. 91, Springer-Verlag, Berlin, 1983.

[Bet] Betsakos, D.: On conformal capacity and Teichmüller's modulus problem in space. - J. Analyse Math. 79 (1999), 201–214.

[BV] Betsakos, D. and M. Vuorinen: Estimates for conformal capacity. - Constr. Approx. 16 (2000), 589–602.

[BB] Borwein, J. M. and P. B. Borwein: Pi and the AGM. - Wiley, New York, 1987.

[Ca] Caraman, P.: On the equivalence of the definitions of the n-dimensional quasiconformal homeomorphisms (QCfH). - Rev. Roumaine Math. Pures Appl. 12 (1967), 889–943.

[DS] David, G. and S. Semmes: Analysis of and on Uniformly Rectifiable Sets. - Mathematical Surveys and Monographs 38 (1993), American Mathematical Society, Providence, RI.

[Fe] Ferrand, J.: Conformal capacities and extremal metrics. - Pacific J. Math. (180) 1 (1997), 41–49.

[Ge1] Gehring, F.W.: Symmetrization of rings in space. - Trans. Amer. Math. Soc. 101 (1961), 499–519.

[Ge2] Gehring, F.W.: Rings and quasiconformal mappings in space. - Trans. Amer. Math. Soc. 103 (1962), 353–393.

[GM] Gehring, F.W. and O. Martio: Quasiextremal distance domains and extension of quasiconformal mappings. - J. Analyse Math. 45 (1985), 181–206.

[GLM] Granlund, S., P. Lindqvist, and O. Martio: F-harmonic measure in space. - Ann. Acad. Sci. Fenn. Ser A I Math 7 (1982), 233–247.

[HKM] Heinonen, J., T. Kilpeläinen, and O. Martio: Nonlinear Potential Theory of Degenerate Elliptic Equations. - Clarendon Press, Oxford, 1993.

[JV] Järvi, P. and M. Vuorinen: Uniformly perfect sets and quasiregular mappings. - J. London Math. Soc. (2) 54 (1996), 515–529.

[LeVu] Lehtinen, M. and M. Vuorinen: On Teichmüller's modulus problem in the plane. - Rev. Roumaine Math. Pures Appl. 33 (1988), 1–2, 97–106.

[LeVi] Lehto, O. and K. I. Virtanen: Quasiconformal mappings in the plane. - Die Grundlehren der math. Wissenschaften Vol. 126, 2nd ed., Springer-Verlag, Berlin-Heidelberg-New York, 1973.

[LF] Lelong-Ferrand, J.: Invariants conformes globaux sur les varietes riemanniennes. - J. Differential Geom. 8 (1973), 487–510.

[Mar1] Martio, O.: Equicontinuity theorem with an application to variational integrals. - Duke Math. J. 42 (1975), 569–581.

[Mar2] Martio, O.: Capacity and measure densities. - Ann. Acad. Sci. Fenn. Ser. A I Math. 4 (1978/1979), 109–118.

[MS] MARTIO, O. and J. SARVAS: Density Conditions in the n-Capacity. - Indiana Univ. Math. J. 26 (1977), 761–776.

[Mat] MATTILA, P.: Geometry of Sets and Measures in Euclidean Spaces. - Cambridge Univ. Press, Cambridge 1995.

[MR] MATTILA, P. and S. RICKMAN: Averages of the counting function of a quasiregular mapping. - Acta Math. 143 (1979), 273–305.

[Re1] RESHETNYAK, YU. G.: Spatial mappings with bounded distortion. - Izdat. "Nauka", Sibirsk. Otdelenie, Novosibirsk, 1982. (Russian)

[Re2] RESHETNYAK, YU. G.: Space mappings with bounded distortion. - Translations of Mathematical Monographs 73, American Mathematical Society, Providence, 1989.

[Ri] RICKMAN, S.: Quasiregular Mappings. - Ergeb. Math. Grenzgeb. (3), Vol. 26, Springer-Verlag, Berlin, 1993.

[Sa1] SARVAS, J.: Symmetrization of condensers in n-space. - Ann. Acad. Sci. Fenn. Ser. A I 522 (1972), 1–44.

[Sa2] SARVAS, J.: The Hausdorff dimension of the branch set of a quasiregular mapping. - Ann. Acad. Sci. Fenn. Ser. A I Math. 1 (1975), 297–307.

[Se] SEITTENRANTA, P.: Möbius-invariant metrics. - Math. Proc. Camb. Phil. Soc. 125 (1999), 511–533.

[SV] SOLYNIN, A. YU. and M. VUORINEN: Extremal problems and symmetrization for plane ring domains. - Trans. Amer. Math. Soc. 348 (1996), 4095–4112.

[Su] SUGAWA, T.: Uniformly perfect sets – analytic and geometric aspects –. - Sugaku 53(4) (2001), 387–402. (Japanese, to appear in English.)

[Te] TEICHMÜLLER, O.: Untersuchungen über konforme und quasikonforme Abbildung. - Deutsche Math. 3 (1938), 621–678.

[Vä1] VÄISÄLÄ, J.: On quasiconformal mappings in space. - Ann. Acad. Sci. Fenn. Ser. A I 298 (1961), 1–36.

[Vä2] VÄISÄLÄ, J.: Lectures on n-Dimensional Quasiconformal Mappings. - Lecture Notes in Math. 229, Springer-Verlag, Berlin-Heidelberg-New York, 1971.

[Vä3] VÄISÄLÄ, J.: Capacity and measure. - Michigan Math. J. 22 (1975), 1–3.

[Vu1] VUORINEN, M.: Exceptional sets and boundary behavior of quasiregular mappings in n-space. - Ann. Acad. Sci. Fenn. Ser. A I Math. Dissertationes 11 (1976), 1–44.

[Vu2] VUORINEN, M.: Lower bounds for the moduli of path families with applications to non-tangential limits of quasiconformal mappings. - Ann. Acad. Sci. Fenn. Ser. A I Math. 4 (1978/1979), 279–291.

[Vu3] VUORINEN, M.: Conformal invariants and quasiregular mappings. - J. Analyse Math. 45 (1985), 69–115.

[Vu4] VUORINEN, M.: Conformal Geometry and Quasiregular Mappings. - Lecture Notes in Math. 1319, Springer-Verlag, Berlin-New York, 1988.

[Vu5] VUORINEN, M.: Conformally invariant extremal problems and quasiconformal maps. - Quart. J. Math. Oxford (2), 43 (1992), 501–514.

[Wa] WALLIN, H.: Metrical characterization of conformal capacity zero. - J. Math. Anal. Appl. 58 (1977), 298–311.

Department of Mathematics
P.O.Box 4 (Yliopistonkatu 5)
FIN–00014 University of Helsinki
Finland
E-mail: `ville.heikkala@helsinki.fi`